**C. Wright Mills and the motorcycle he used
to commute to Columbia University in the 1950s**
Photo courtesy of Yaroslava Mills

Beyond Sociology's Tower of Babel

SOCIOLOGICAL IMAGINATION
AND STRUCTURAL CHANGE
An Aldine de Gruyter Series of Texts and Monographs

SERIES EDITORS
Bernard Phillips, *Professor Emeritus, Boston University*
Harold Kincaid, *University of Alabama, Birmingham*

Lawrence Busch
THE ECLIPSE OF MORALITY
Science, State, and Market

Leo d'Anjou
SOCIAL MOVEMENTS AND SOCIAL CHANGE
The First Abolition Campaign Revisited

Frank Hearn
MORAL ORDER AND SOCIAL DISORDER
The American Search for Civil Society

David R. Maines
THE FAULTLINE OF CONSCIOUSNESS
A View of Interactionism in Sociology

Pierre Moessinger
THE PARADOX OF SOCIAL ORDER
Linking Psychology and Sociology

Bernard Phillips
BEYOND SOCIOLOGY'S TOWER OF BABEL
Reconstructing the Scientific Method

Beyond Sociology's Tower of Babel

Reconstructing the Scientific Method

Bernard Phillips

ALDINE DE GRUYTER
New York

ABOUT THE AUTHOR

Bernard Phillips was introduced to sociology at Columbia University by C. Wright Mills. A former Professor of Sociology at Boston University, cofounder of the ASA Section on Sociological Practice and founder of the Sociological Imagination Group, his publications emphasize methodology and theory.

Copyright © 2001 Walter de Gruyter, Inc., New York

ALDINE DE GRUYTER
A division of Walter de Gruyter, Inc.
200 Saw Mill River Road
Hawthorne, New York 10532

This publication is printed on acid free paper ⊗

Library of Congress Cataloging-in-Publication Data

Phillips, Bernard S.
    Beyond Sociology's Tower of Babel : reconstructing the scientific method / Bernard Phillips.
        p. cm. — (Sociological imagination & structural change)
    Includes bibliographical references and index.
    ISBN 0-202-30665-8 (cloth : alk. paper) — ISBN 0-202-30666-6 (pbk. : alk. paper)
    1. Sociology—Methodology. 2. Science—Methodology. I. Title. II. Sociological imagination and structural change

HM511 .P54 2001
301'.01—dc21

                                                                2001037247

Manufactured in the United States of America

10 9 8 7 6 5 4 3 2 1

*In memory of C. Wright Mills
who gave us a vision for sociology's future*

Contents

Preface

C. Wright Mills, who turned me as a student at Columbia College in 1950 from medicine to sociology, embodied Auguste Comte's Enlightenment dream as to the promise of sociology. If the biophysical sciences had created the basis for problem-solving technologies now known as the industrial revolution, surely the discipline of sociology could manage to do the same for the enormous social problems facing society. If the scientific method had been the chief instrument for transforming societies from the seventeenth century onward, why couldn't that same instrument help us to direct that transformation toward ends that human beings desire? The attention given to Mills by the discipline, despite the fact that he was a loner who wrote mainly for a popular audience, is illustrated by the rating given to *The Sociological Imagination* in 1998 by the members of the International Sociological Association. It achieved the second highest rating among books published in the twentieth century considered to be the most influential for sociologists, preceded by Weber's *Economy and Society* and followed by Merton's *Social Theory and Social Structure*, Weber's *The Protestant Ethic and the Spirit of Capitalism*, Berger's and Luckman's *The Social Construction of Reality*, Bourdieu's *Distinction*, Elias' *The Civilizing Process*, Habermas' *The Theory of Communicative Action*, Parsons' *The Structure of Social Action* and Goffman's *The Presentation of Self in Everyday Life*.

Despite that rating along with the widespread lip service given to the idea of "the sociological imagination," Mills's methodological achievements in that book and elsewhere have been barely noticed relative to his contributions to political sociology. He gave us a vision of a sociology that would dare to define the most fundamental problems facing the human race as research problems. I recall a day in the spring of 1958 when we were both on a plane to Champaign-Urbana where he was to deliver a lecture based on his book, *The Causes of World War III*. In addition to his defining basic research problems, the breadth of the approach he advised for all sociologists, and others as well, is suggested by this familiar quote:

> . . . that imagination is the capacity to shift from one perspective to another—from the political to the psychological; from examination of a single family to comparative assessments of the national budgets of the world . . . [1959: 7].

And there is also his reflexive orientation, illustrated in his Appendix—
"On Intellectual Craftsmanship"—with these words: "The most admirable
thinkers within the scholarly community you have chosen to join do not
split their work from their lives." This was an orientation strengthened in
1970 by Gouldner's call for a "reflexive sociology" in *The Coming Crisis of
Western Sociology*.

What is the state of modern sociology with respect to Mills's orientation
to basic problems, to intellectual breadth and to a reflexive orientation? In
the early 1940s while still a graduate student at the University of Wisconsin, Mills analyzed some fifty textbooks on social problems in order to
learn about "the style of reflection and the social-historical basis of American sociology":

> The level of abstraction which characterizes these texts is so low that often
> they seem to be empirically confused for lack of abstraction to knit them to-
> gether. They display bodies of meagerly connected facts, ranging from rape
> in rural districts to public housing, and intellectually sanction this low level
> of abstraction . . . Collecting and dealing in a fragmentary way with scattered
> problems and facts of milieux, these books are not focused on larger stratifi-
> cations or upon structured wholes [1943: 166].

Mills's criticism could just as easily apply to contemporary sociology. We
now have some forty Sections of the American Sociological Association
with their "meagerly connected facts." And a basis for this state of affairs is
still a "low level of abstraction," given our orientation to Merton's "theories
of the middle range." The problems we tend to define for study are no more
fundamental than "rape in rural districts" and "public housing." And despite Mills and Gouldner, our research is characterized by little reflexivity.

It appears that at this time in history, we have experienced a century of
sociology's failure to achieve the rapid cumulative development sought by
Comte, Mills, Gouldner, and the rest of us, and this has been coupled with
a century of awesome and apparently escalating problems in modern society. Given this state of affairs, it is no wonder that almost all of us have
turned to "the falsification of memory" and "the technique of particularization," procedures described by Vidich and Bensman in their 1960 study
of Springdalers. We have largely succeeded in burying our Enlightenment
dreams for the promise of sociology by immersing ourselves within highly
specialized areas of study. We behave much like the Springdaler who "instead of entertaining the youthful dream of a 500-acre farm, entertains the
plan to buy a home freezer by the fall." The result for us appears to be much
like that for the Springdalers:

> Because they do not recognize their defeat, they are not defeated. The com-
> promises, the self-deception and the self-avoidance are mechanisms which

work; for, in operating on the basis of contradictory, illogical and conflicting assumptions, they are able to cope in their day-to-day lives with their immediate problems in a way that permits some degree of satisfaction, recognition and achievement . . . [1960: 320].

Yet as we examine our situation in the twenty-first century, we contemporary sociologists find ourselves in a position to open up to those falsified memories and to challenge our techniques of particularization by building on Mills's ideas, such as his emphasis on basic problems. For example, the very fact that modern problems appear to be increasing—such as the escalation of technologies for delivering weapons of mass destruction—makes it ever more difficult for us to continue to bury our heads in the sand. It is becoming more and more obvious to us sociologists that we have failed to give society the platform of knowledge it needs as a basis for constructing social technologies in all institutions that can confront our complex social problems. If we look outside the discipline we find little credibility given to us as scientists, and we are learning to see ever more clearly the contradiction between our scientific ideals and what we have in fact achieved within the discipline. Instead of sociological knowledge based on the full range of our findings, we have separate pieces of knowledge located within the diverse areas of the discipline. Instead of knowledge that is rapidly cumulating, we have fads and fashions in the ideas and terms we use with relatively little cumulative development. We even have questions raised by some postmodernists, as well as others, as to the possibility of any "scientific method" that can be applied to human behavior.

In addition to this increasing attention to fundamental problems outside and inside of the discipline, we are now in a much better position to follow Mills's lead of shuttling up and down language's ladder of abstraction, giving us increasing ability to integrate our "bodies of meagerly connected facts, ranging from rape in rural districts to public housing." For example, the very fragmentation of sociology into forty sections has yielded greater understanding of the enormous complexity of human behavior, suggesting that pessimism and cynicism concerning the discipline's achievements and potential is premature. Furthermore, a new sociological understanding of the limitations of positivism's quantitative one-sidedness opens up a much broader approach to the scientific method. Abstract concepts and theories have come to the fore as essential ingredients for developing a science of sociology, just as they have proved to be essential for the biophysical sciences. The contemporary philosophy of science and social science points up alternatives to positivistic assumptions, such as the inevitability of being carried to truth by the scientific method, the importance of isolated hypotheses, and the centrality of exact prediction. Instead, we emerge with a web orientation to the scientific method, where no proposition is seen in isolation from all others, and where, following Mills, we shuttle up and down language's ladder of abstraction.

It is also possible to go beyond Mills's and Gouldner's ideas about reflexivity. To illustrate, we might begin with Kuhn's view of scientific revolutions as stemming from awareness of contradictions within an existing scientific paradigm that are resolved within an alternative one. By extending his idea of scientific paradigms to cultural paradigms, his thesis implies that cultural revolutions also require awareness of existing contradictions that are resolved within an alternative cultural paradigm. Reflexivity points us exactly in this direction, examining our work and life with an eye toward uncovering contradictions, and coming up with an alternative cultural paradigm where they could be resolved. Such reflexivity need not be limited to a vague metaphor. Instead, the concept of "cultural paradigm" can become a highly systematic one when linked to other abstract concepts within the reconstructed scientific method suggested by Mills's notion of shuttling up and down the ladder of abstraction. From this perspective, we can proceed systematically to question—reflexively—our taken-for-granted assumptions, following the direction suggested metaphorically by Gouldner:

> The historical mission of a Reflexive Sociology . . . would be to *transform* the sociologist, to penetrate deeply into his daily life and work . . . and to raise the sociologist's self-awareness to a new historical level [1970: 489].

The difficulties involved in reconstructing our present approach to the scientific method appear to be extraordinary, for that approach is nested within our cultural paradigm. Just as we have relatively isolated sections within the American Sociological Association, so do we have institutions within society along with organizations within the economy and occupations within organizations that have little contact with one another. Granting that a reflexive approach would raise this nested relationship up to full view, we would then be presented with an enormous contradiction between scientific ideals and practices that are supported by our fundamental way of life. Just as in the case of the Springdalers, we would be presented with youthful ideals that we have failed to reach. Under these circumstances, what would prevent us from reverting to our present cultural paradigm that gives us procedures for the falsification of memory through techniques of particularization? By so doing, we would at least be able to obtain "some degree of recognition, satisfaction and achievement." This is precisely the difficulty I face as author of this book. I am asking readers to bring to the surface taken-for-granted assumptions about the scientific method, and to examine their departure from scientific ideals, a task that also will show the departure of our cultural paradigm from our basic cultural values.

Yet if we follow Kuhn's argument further—an argument strengthened by much that sociologists have learned about society—we find that awareness of such contradictions can yield changes in a scientific paradigm. If a new paradigm is developed within which those contradictions are resolved, then we can expect a shift to that paradigm, and this is exactly the approach I take in the following chapters. I outline new scientific and cultural paradigms, and sketch their potential for resolving contradictions within the old ones. I do this not because I believe that I have succeeded in demolishing our present scientific and cultural paradigms. Rather, I believe it essential for a reader to be presented with these alternatives in order to consider the possibility of such drastic and all-encompassing changes. Indeed, in my own view, I do no more than open a door to the possibility of such shifts in our approach to the scientific method as well as modern culture. Fortunately, from my own perspective, I am not alone in this approach, for it is the basis for the work of a group made up largely of sociologists, the sociological imagination group, introduced in a web site at <www.uab.edu/ethicscenter/SI.html>. We ask other sociologists and social scientists to join us in our present efforts to test the utility of this approach to the scientific method. We feel that it is worthwhile to abandon our present degree of "recognition, satisfaction and achievement," given the urgent problems in sociology and society.

In Part I, "The Scientific Method: Bureaucratic and Interactive Paradigms," I present in Chapter 1 a direction for reconstructing the scientific method. It builds on C. Wright Mills's work and is oriented to shuttling far up and down language's ladder of abstraction, in contrast with "grand theory" or "abstracted empiricism" which emphasize the top or the bottom, respectively, of that ladder. His analysis includes his vision of the centrality of developing a sociological imagination, and here he points beyond sociology to society as a whole, just as his books were written for a popular as well as an academic audience. Chapter 2 uses Kuhn's analysis as a basis for taking up our cultural paradigm as well as our sociological paradigm, with the latter nested within the former just as any given epistemology is nested within a metaphysics. This chapter carries further Chapter 1's analysis of the contrast between bureaucratic and interactive cultural and scientific paradigms. We look to the ideas of John Dewey, Thomas Kuhn, and Harold Kincaid for philosophical and historical insights. David Snow and Thomas Scheff—devoting particular attention to sociology's methods and theory—give us further understanding of sociology's present situation as well as its future possibilities. Overall, Part I presents an outline of a reconstructed scientific method, taking into account some of its implications for modern society.

Part II, "Illustrating the Web Approach to the Scientific Method," aims to exemplify both the substantive fruitfulness and the applied implications of that approach. Yet the focus of Part II remains on the web approach to the scientific method, and not on the validity of the few substantive and applied illustrations presented. Chapter 3 centers on anomie, alienation, social stratification, and relative deprivation, all aspects of one fundamental social problem that might be called "the invisible crisis of modern society." The web approach to the scientific method may prove to yield understanding of very broad and basic problems, as well as those of limited scope, that encompass the full range of substantive questions within the discipline. Whereas Chapter 3 centers on the utility of the web approach for understanding problems, Chapter 4 focuses on its utility for moving toward solutions. There we begin with a general examination of what it would take to change both sociology's research paradigm and the cultural paradigm within modern society. We follow this with specific analyses of three illustrations: revolutions in general, the Gandhian technique of *satyagraha*, and change in a two-person social relationship. Both parts of Chapter 4 are oriented to the problem of how social structures can be altered so as to yield a higher degree social interaction.

Part III, "Some Implications for Sociology" looks to several implications of the foregoing chapters for the discipline of sociology. In Chapter 5, we center on Gouldner's idea of a reflexive sociology and point toward procedures that would carry forward Gouldner's vision. In particular, we take up a number of approaches to social change along with educational procedures. In both cases, we examine the impact of reflexivity. Thus, for example, we look to what a reflexive approach would yield for culture lag theory. We also carry forward the reflexive implications of some of the educational ideas of Dewey, Freire, Illich, Pecotche, and Gandhi. Chapter 6 centers on language as a key thread tying together earlier chapters. In an initial section, "Strengthening Linguistic Tools," we introduce additional concepts to help us integrate earlier material, and we carry further our understanding of causal-loop diagrams. Also, we summarize two allegories—*Nineteen Eighty-Four* and *The Languages of Pao*—for penetrating the nature of language. Then, in a second section, we apply these linguistic tools to earlier chapters, centering on the problem of emotional repression and the possibilities for emotional expression.

If the arguments in this book for the problematic nature of our present interpretation of the scientific method prove to be credible, and if the alternative interpretation sketched here proves to be fruitful, then the implications of those arguments and that interpretation extend far beyond the few that are examined in Part III. For example, the conclusions drawn for every single social science study that has ever been conducted would be open to reinterpretation. This would result from past failures to take into account systematically the enormous complexity involved within any given instance of human behavior. Our present piecemeal and specialized approach assumes implicitly that the pieces of the human jigsaw puzzle can at some point be put together so as to yield a coherent picture of human behavior. Yet if each piece is itself deficient, then no coherent picture emerges when we attempt to put the pieces together. Shifting from this metaphor to research procedures presently in use, one example has to do with the impact of the investigator on the investigation at every stage of the research process. These "investigator effects" are not taken into account by our non-reflexive approach to the scientific method, and we continue to publish studies which almost invariably include no information about such effects despite occasional questions raised as to the unscientific nature of such practices. This is much like a trial where neither the prosecutor nor the defense attorney is allowed to cross-examine witnesses yet where we expect an accurate verdict.

Carrying this implication one step further, if all of our conclusions from the social sciences become suspect, then so does the worth of all of the actions based on those conclusions which we have performed as individuals and societies. This includes past and present decisions made within every one of our institutions, and this is not limited to the relatively few decisions based on proposals by social scientists. They extend to the subtle influences of the social sciences as a result of their location within our formal and informal educational systems at all levels. All of this is also implied by the limitations within the cultural paradigm of modern society, a paradigm that encompasses the scientific paradigm governing our research procedures. In one sense, this far-reaching critique of modern society is not a new idea, since much of postmodernist literature suggests the importance of deconstructing present assumptions within all aspects of society. Yet what is new is the acceptance of such deconstruction *coupled with* an alternative approach to the scientific method that is optimistic in its assessment of human possibilities, and that promises to resolve fundamental contradictions within our scientific and cultural paradigms.

That alternative approach also gives the social scientist a special role within contemporary society, as illustrated by Gouldner's vision of the future of the social sciences:

> . . . At decisive points the ordinary language and conventional understandings fail and must be transcended. It is essentially the task of the social sciences, more generally, to create new and "extraordinary" languages, to help men learn to speak them, and to mediate between the deficient understandings of ordinary language and the different and liberating perspectives of the extraordinary languages of social theory. . . . To say social theorists are concept-creators means that they are not merely in the *knowledge*-creating business, but also in the *language*-reform and language-creating business. In other words, they are from the beginning involved in creating a new *culture* [Gouldner, 1972: 16].

Following Gouldner's vision, perhaps the twenty-first century will not come to be seen as the century that witnessed an acceleration of the catastrophes of the twentieth century. Instead, perhaps it will come to be seen as the age of the social sciences, where the Enlightenment visions of Comte, Mills, Gouldner, and the rest of us for a society able to confront its fundamental problems became a reality for the first time in human history.

In a book where I attempt to build bridges linking sociological knowledge, everything I have ever read inside and outside of the discipline has influenced me. In particular, I am indebted to every individual cited in the following chapters for making this book possible. Earlier ideas from Marx, Simmel, Durkheim, Weber, Nietzsche, and Korzybski have been particularly influential, as is the case for more recent ideas from Mills, Gouldner, Kuhn, and Williams. I owe a great deal to Harold Kincaid, whose philosophical ideas helped to start me on this journey, to Tom Scheff, whose part/whole methodology and publications helped to broaden my orientation, to Richard Koffler for his faith in my ideas, and to Louie Johnston for his unflagging enthusiasm. I want to thank those who read parts of this manuscript at one stage or another, or who encouraged me to proceed, including Dave Asavanond, Steve Baran, Stu Bennett, Larry Busch, Lee Cass, David Christner, Hank Everett, Joe Feagin, Sandy Klein, Marty Kozloff, Jack Levin, Felice Levine, Tony Levy, John and Joanne Livingstone, Clem Malin, Marvin Nadel, David Phillips, Seymour and Phyllis Pustilnik, John Rice, Dave Stearns, and Emek Tanay. I have learned much from all those who are a part of the sociological imagination network, with special thanks to Dave Britt, Tom Conroy, Dick Edgar, John Hall, Matt Hoover, Joe Hopper, Chanoch Jacobsen, Jim Kimberly, Richard Lachmann, Lauren Langman, Donald Levine, Bronwen Lichtenstein, Dave Maines, John Malarkey, Stjepan Mestrovic, Alfonso Morales, Gil Musolf, Joe Perry, Gary Reed, Jay Weinstein, Doris Wilkinson, and Andy Ziner.

PART I

THE SCIENTIFIC METHOD: BUREAUCRATIC AND INTERACTIVE PARADIGMS

It is in Genesis that God punishes the human race for attempting to build "a town and a tower with its top reaching heaven":

> Now Yahweh came down to see the town and the tower that the sons of man had built. "So they are all a single people with a single language!" said Yahweh. "This is but the start of their undertakings! There will be nothing too hard for them to do. Come, let us go down and confuse their language on the spot so that they can no longer understand one another." Yahweh scattered them thence over the whole face of the earth, and they stopped building the town. It was named Babel therefore, because there Yahweh confused the language of the whole earth. It was from there that Yahweh scattered them over the whole face of the earth [Genesis 11: 5–9; Jerusalem Bible 1966: 14].

From the perspective of the Old Testament, the tower of Babel becomes a metaphor for the division of the human race into groups unable to communicate with one another. Applying that metaphor to contemporary sociology, we appear to have achieved a more subtle procedure than speaking languages from different cultures. We have learned to speak the languages of different subcultures within our discipline. Unless an individual learns the language of a given field by becoming familiar with its literature, he or she will remain unable to communicate with others in that field.

Part I is about an approach to the scientific method that aims at building bridges across those subcultures or fields of sociology, changing our tower of Babel into a discipline where we can all gain from learning to follow the scientific ideal of communicating with one another. Following the work of Thomas Kuhn, we should not underestimate the difficulties confronting any major shift in our discipline. This is particularly true when a scientific paradigm is itself nested within a cultural paradigm. Yet also following Kuhn, scientific revolutions can indeed occur when a discipline becomes aware not only of its fundamental contradictions but also of a direction for

1

resolving them. In Chapter 1 we sketch a contrast between our present interpretation of the scientific method and an alternative interpretation, with the former labeled "bureaucratic" and the latter "interactive." Granting the achievements we have made with the aid of our past interpretation, the method appears to be unable to cope with the enormous complexity of human behavior. Our interactive interpretation or "web approach" aims to take fully into account that complexity. In Chapter 2 we pursue the paradigmatic basis for making a fundamental change to that web approach. For example, we examine the nature of our cultural paradigm as well as a cultural alternative. And we also look to an approach to sociological theory that can enable us to follow that cultural alternative.

1

Sociology and the Scientific Method

The work of C. Wright Mills, someone whom I was fortunate enough to know personally and who envisioned both what our discipline lacks and what it might proceed to achieve, plays an important role in this chapter. Mills is known largely for his contributions to political sociology as well as his metaphor of "the sociological imagination," yet it is time that we begin to take seriously what he contributed to our understanding of the scientific method in sociology. In the first section, "Problems with Bureaucratic Science," I sketch—within the context of other materials—his profound critique not just of the sociology of his own times but also of how we are presently going about our business. To complete the picture he drew, I also bring in aspects of what we have learned since his time from the history and philosophy of science as well as from sociology. This critique of bureaucratic science constitutes the basis, in the second and final part of this chapter, for a more systematic presentation of the approach to the scientific method which I believe we desperately need for substantive progress and which society urgently requires. Mills somehow succeeded in giving voice to the aspirations for sociology that continue to lie buried under layers of cynicism and pessimism within the rest of us sociologists, waiting to take wing.

PROBLEMS WITH BUREAUCRATIC SCIENCE

Many of us are familiar with Mills's concept of the sociological imagination:

> ... that imagination is the capacity to shift from one perspective to another—from the political to the psychological; from examination of a single family to comparative assessments of the national budgets of the world; from the theological school to the military establishment; from considerations of an oil industry to studies of contemporary poetry. (1959:7)

Mills reveals the same breadth of vision that characterized the classical sociologists and is so central to the present ideals of the discipline (see, for example, Horowitz 1983; Chasin 1990). He saw that kind of breadth as essential for fulfilling the Enlightenment "promise of sociology." Although he never developed a systematic direction for just how sociologists should proceed to employ the scientific method, the body of his work suggests five components: (1) We should not shirk from addressing absolutely fundamental problems within society. (2) We should move far up language's ladder of abstraction so as to utilize very abstract concepts. (3) We should come far down that ladder so as to examine the concrete evidence that bears on our ideas. (4) We should work to integrate our knowledge so that our approach is broad enough to enable us—as indicated in the above quote—to shift from one perspective to another. (5) We should develop ourselves as individuals with the ability to think in this broad way, developing a "sociological imagination" that suggests a new vision of society.

Orientation to Problems

Mills's orientation to problems is illustrated by a body of work that included an examination of the power of elites in subverting democratic ideals (1948; 1956), interest in the alienation of "the new middle class" (1951), concerns about the coming of World War III (1958), and an analysis of the personal troubles of the individual in modern society (1959).

We can begin to understand his zest for conflict—expressed in his relationships to sociological colleagues—from a story he told a class about an encounter with Eisenhower, then president of Columbia University. When Eisenhower walked into his classroom one day unannounced and quietly took a seat in the back row, Mills unhesitatingly and without missing a beat altered his lecture, presenting a systematic plan for a violent revolution against the government of the United States. The class would operate as a key cell in directing the course of the revolution, which he claimed was absolutely essential because any other means would be opposed by the ruling class. Eisenhower sat stonily silent as Mills proceeded to state just what would be required if the government were in fact to be overthrown. Finally, Eisenhower stood up and quickly walked out, with Mills never hearing from him afterwards. Mills told this story with a broad smile on his face, having no particular ax to grind against Eisenhower but simply as a joke he managed to play on the powers that be within the university as well as within society as a whole.

We can understand more fully Mills's interest in avoiding trivial problems from a study he completed during World War II while still a graduate student at the University of Wisconsin. He analyzed some fifty textbooks on social problems in order to learn about "the style of reflection and the social-historical basis of American sociology," a rather tall order even for an eminent professor. These texts generally centered on how the individual might "adjust" to society in order to solve his or her problems rather than on how society might change in response to those problems:

Use of "adjustment" accepts the goals and the means of smaller community milieux. At the most, writers using these terms suggest techniques or means believed to be less disruptive than others to attain the goals that are given. They do not typically consider whether or not certain groups or individuals caught in economically underprivileged situations can possibly obtain the current goals without drastic shifts in the basic institutions which channel and promote them. The idea of adjustment seems to be most directly applicable to a social scene in which, on the one hand, there is a society and, on the other, an individual immigrant. The immigrant then "adjusts" to the new environment. (1943:179–80).

For Mills, the idea of adjustment works to foreclose the aspirations of the immigrant. An alternative would be for the immigrant to retain those high aspirations and for society to undergo "drastic shifts in the basic institutions which channel and promote them."

Mills also cites the emphasis of these texts on the idea of "cultural lag," based on the work of Ogburn with his idea that "adaptive" or "nonmaterial culture" lags behind "material culture":

The model in which institutions lag behind technology and science involves a positive evaluation of natural science and of orderly progressive change. Loosely, it derives from a liberal continuation of the enlightenment with its full rationalism, its messianic and now politically naive admiration of physical science as a kind of thinking and activity, and with its concept of time as progress. (Mills 1943: 177)

This use of the idea of cultural lag is analogous to an emphasis on the adjustment of the individual. Not only must the individual adapt to existing norms of society but the institutions of society must adapt to the physical and biological technologies built on the continuing development of the physical and biological sciences, for continuing "progress" supposedly depends on such adaptation. Yet we can construct alternatives to this approach, just as Mills saw alternatives to accepting the goal of the adjustment of the immigrant as one that sociologists should adopt. Society need not bow down to physical and biological technologies involving "material culture." Rather, it is possible for us to create the kind of society in which those technologies move in directions that strengthen "nonmaterial culture."

Mills's approach to the scientific method within sociology preceded Thomas Kuhn's (1962, 1977, 1992) analysis of scientific revolutions within the physical and biological sciences by several years, yet the two orientations are similar in their view of the forces blocking us from confronting basic problems. Kuhn's argument was based on an intuitive application of the sociology of knowledge. He saw a community of scientists as being swayed not just by evidence alone but also by such factors as tradition, social hierarchy, and the personality of the scientist. For Kuhn, the problem of achieving a scientific revolution is a massive one, for the very subculture of a science or its "scientific paradigm" must be challenged and not just particular studies. Mills is also concerned with the subculture of a science. For example, he suggests that "academic departmentalization may well have been instrumental in atomizing the problems which they [the authors] have addressed" (1943:166), thus affecting the failure to confront the large question of changing institutions. Mills also suggests that those authors came from similar backgrounds, shared common perspectives, and thus tended to conform to relatively conservative norms when it came to any question of fundamental changes in society. As for hierarchy, Mills points to the relationship between teacher and student or author of a text and students who read the text. The result is an emphasis on the systematization of existing ideas rather than any questioning of those ideas or attempting to discover new ones.

If we take Kuhn's analysis one step further, then we can discover problems that are even more fundamental than subcultural ones, namely, cultural ones. Extrapolating Kuhn, research can succeed in confronting contradictions within the researcher's cultural paradigm. For an example we turn to Friedrich Nietzsche's *The Gay Science:*

> The greatest recent event—that "God is dead," that the belief in the Christian god has become unbelievable—is already beginning to cast its first shadows over Europe. . . . How much must collapse now that this faith has been undermined . . . for example, the whole of our European morality. . . . We philosophers and "free spirits" feel, when we hear the news that "the old god is dead," as if a new dawn shone on us . . . at long last our ships may venture out again, venture out to face any danger; all the daring of the lover of knowledge is permitted again. ([1887] 1974: 279-280).

Mills no less than Nietzsche was most interested in investigating major transformations of society. He believed that "The existence of mass estrangement among workers, anxiety among professionals, and anomie among middle sectors invalidated the 'modern' period," and he suggested that the new epoch that was dawning might be labeled the "postmodern era," apparently the first scholar to use that term. And just as Kuhn wrote of contradictions that come to light within the old paradigm, Mills wrote of the ending of an epoch: "When what is happening in the social world as well as what is widely felt and widely thought can no longer be satisfactorily explained by the received principles, then an epoch is ending and a new one needs to be defined" (1960; quoted in Horowitz 1983:323, 327).

Moving Up Language's Ladder of Abstraction

Yet just how are major problems, such as our present failure to understand "what is happening in the social world as well as what is widely felt and widely thought," to be addressed? Or using Nietzsche's example, just how is the "death of God" to be understood and confronted? Mills's analysis suggests an answer when he employs the concept of "anomie" and when his usage of "estrangement among workers" and "anxiety among professionals" implies the concept of "alienation." It is, then, sociology's abstract concepts that come to function as a basic part of the scientific method within the discipline. Alvin Gouldner, in a reply to a review of his *The Coming Crisis of Western Sociology* (1970), explains more clearly what Mills was implying:

> At decisive points the ordinary language and conventional understandings fall and must be transcended. It is essentially the task of the social sciences, more generally, to create new and "extraordinary" languages, to help men learn to speak them, and to mediate between the deficient understandings of ordinary language and the different and liberating perspectives of the extraordinary languages of social theory. . . . To say social theorists are concept-creators means that they are not merely in the *knowledge*-creating business, but also in the *language*-reform and language-creating business. In other words, they are from the beginning involved in creating a new *culture*. (Gouldner 1972:16; emphasis in original)

Concepts like anomie and alienation are, then, the sociologist's most powerful tool for understanding the fundamental problems of society.

From this perspective we can draw an analogy between ordinary language and the language of sociology, looking to both similarities and differences. On the one hand, ordinary and sociological concepts are tools for understanding our world and addressing problems, and both can be employed within sentences or propositions that state the nature of the world and how we might proceed to solve problems. Also, both are abstract to at least some degree, being at least one stage removed from our nonverbal experiences with phenomena: they are linguistic and reflect on experience, as distinct from the tools used by other forms of life. On the other hand the linguistic tools of the sociologist carry along with them the weight of sociological knowledge, and such knowledge can be most useful in understanding the world and addressing our problems. In addition, the concepts of sociology—like anomie and alienation—generally are more abstract than our everyday concepts, which tend to stay closer to whatever we experience concretely. As a result, sociological concepts tend to cover far more ground. When we speak of anomie, for example, we can refer to contradictions within any culture, past or present. Yet if the sociologist remains unaware of the importance of these differences then it will be all too easy to rely far too much on ordinary usages, using sociological language only in passing. Also, the sociologist will fail to define sociological concepts in a sufficiently abstract way, thus losing generality.

Mills wrote about such deficiencies within sociological usage in his analysis of texts on social problems:

> The level of abstraction which characterizes these texts is so low that often they seem to be empirically confused for lack of abstraction to knit them together. They display bodies of meagerly connected facts, ranging from rape in rural districts to public housing, and intellectually sanction this low level of abstraction. . . . Collecting and dealing in a fragmentary way with scattered problems and facts of milieux, these books are not focused on larger stratifications or upon structured wholes. (1943:166)

Mills suggests that we require abstract concepts like social stratification, concepts general enough to apply to a very wide range of situations, if we are to avoid the "bodies of meagerly connected facts" to be found in those texts. Yet usage of abstract concepts has immediate implications for our approach to problems. If patterns of social stratification are a partial cause of social problems, then solutions will require fundamental changes in society. By avoiding such usages, the sociologist is able to maintain a more conservative stance on addressing social problems. What Mills succeeds in doing is laying bare just what is going on, confronting sociologists with contradictions between their scientific ideals and their actual research procedures. In Kuhn's terms, he alerts us to paradigmatic contradictions within the subculture of sociology as well as within modern culture.

Mills is taking on not just sociologists who center on social problems but also an emphasis within the discipline as a whole, as illustrated by what Merton has called "theories of the middle range":

> Every effort should be made to avoid dwelling upon illustrations drawn from the "more mature" sciences—such as physics and chemistry . . . because their very maturity permits these disciplines to deal fruitfully with abstractions of high order to a degree which, it is submitted, is not yet the case with sociology. (1968:139–40)

Merton's approach appears to be a classic illustration of what he himself called a "self-fulfilling prophecy." By defining sociology as immature and unable to employ "abstractions of high order," we create that very situation of immaturity. Some three decades ago Willer and Webster (1970) launched a profound critique of Merton's approach to the scientific method, basing their argument largely on the philosophy of science (see also Peirce 1955; Hempel 1965; Willer 1967). They maintained that the more developed sciences, versus sociology, construct abstract concepts. For example, there are assertions about "mass" and "specific gravity" in physics, about "bonds" and "valences of molecules" in chemistry, and about "heredity," "natural selection," and "genes" in biology. They argued that sociology's "immaturity" derives in large measure from its failure to use abstract concepts and theory. This approach taken by Willer and Webster has been updated by a variety of analyses pointing in the same direction (see, for example, Phillips 1972, 1979, 1985, 1988, 1990; Lauderdale, McLaughlin, and Oliverio 1990; Wallerstein 1980, 1991, 1998).

Sociology's failure to emphasize concepts at a very high level of abstraction also derives from the relationship between the disciplines of sociology and philosophy coupled with the pragmatic stance of much of American sociology. As a discipline sociology has in large measure distinguished itself from philosophy by emphasizing its empirical stance. That stance has been seen as enabling it to confront the practical problems of everyday life. Mills saw this situation as follows:

> The ideal of practicality, of not being "utopian," operated ... as a polemic against the "philosophy of history" brought into American sociology by men trained in Germany; this polemic implemented the drive to lower levels of abstraction. A view of isolated and immediate problems as the "real" problems may well be characteristic of a society rapidly growing and expanding, as America was in the nineteenth century and, ideologically, in the early twentieth century. . . . The practice of the detailed and complete empiricism of the survey is justified by an epistemology of gross description. (1943:168)

Mills reveals the contradiction between ideals for practical action, on the one hand, and scientific procedures that tend to atomize problems and prevent the sociologist from understanding them. The very tools required for such understanding come to be seen negatively as "philosophical" and are as a result avoided. And the result is the kind of fragmentation of problems that Mills noted in his analysis of textbooks on social problems.

Moving Down the Ladder of Abstraction

Let us bear in mind as we proceed to examine movement down language's ladder of abstraction how intimately it is linked to both commitment to a problem and the use of abstract concepts. It is useful to return to Nietzsche at this point, for he conveys the spirit of science's empirical achievements by contrast with the situation of humanity prior to the rise of science:

It is a profound and fundamental good fortune that scientific discoveries stand up under examination and furnish the basis, again and again, for further discoveries. . . . To lose firm ground for once! To float! To err! To be mad! That was part of the paradise and the debauchery of bygone ages, while our bliss is like that of a man who has suffered shipwreck, climbed ashore, and now stands with both feet on the firm old earth—amazed that it does not waver. ([1887] 1974:111)

Nietzsche here captures science's ability to discover firm *knowledge*—not necessarily absolute truth—which far transcends prescientific *opinions*. It is this ability that is fundamental to the motivation of sociologists in their efforts to move down language's ladder of abstraction to obtain facts. Yet when that motivation becomes isolated from commitments to solving major problems and using abstract concepts in the process, then it yields what Mills criticized as trivial research despite its helping us to plant "both feet on the firm old earth."

Sociologists are able to move down the ladder of abstraction by using procedures that oversimplify the complexity of human behavior. Implicitly, this appears to be an effort to imitate both the simplicity and predictive power of much of the physical sciences, as illustrated by the simple yet powerful formula, $F = ma$, or force = mass times acceleration. Herbert Blumer criticized this approach many years ago:

The objective of variable research is initially to isolate a simple and fixed relation between two variables. . . . This is accomplished by separating the variable from its connection with other variables through their exclusion or neutralization.

A difficulty of this scheme is that the empirical reference of a true sociological variable is not unitary or distinct. When caught in its actual social character, it turns out to be an intricate and inner-moving complex. To illustrate, let me take what seems ostensibly to be a fairly clear-cut variable relation, namely between a birth control program and the birth rate of a given people. . . . For the program of birth control one may choose its time period, or select some reasonable measure such as the number of people visiting birth control clinics. For the birth rate, one merely takes it as it is . . .

Yet, a scrutiny of what the two variables stand for in the life of the group gives us a different picture. Thus, viewing the program of birth control in terms of *how it enters into the lives of people*, we need to note many things such as the literacy of the people, the clarity of the printed information, the manner and extent of its distribution, the social position of the directors of the program and of the personnel, how the personnel act, the character of their instructional talks, the way in which people define attendance at birth control clinics, the expressed views of influential personages with reference to the program, how such persons are regarded, and the nature of the discussions among people with regard to the clinics. (Blumer 1956:688)

Blumer is here criticizing the simplifying assumptions, largely invisible, that lie behind what he calls "analysis of the variable" within quantitative sociology. Granting the importance of moving down language's ladder of abstraction to concrete measurements of particular factors, we sociologists should have learned enough from our research to realize that we cannot learn much by centering on only two variables within a complex context of factors and ignore the rest with some phrase like "other things being equal" or *ceteris paribus*. And we can even add many other factors to Blumer's example, which emphasizes situational factors close to his own symbolic-interactionist perspective. For example, there is the little matter of cultural values and norms as well as patterns of social organization in society as a whole like social stratification and bureaucracy. Most of quantitative methodology pushes aside such considerations, for they would interfere with the tools of measurement that we presently have and that yield definite findings. Those tools build on certain aspects of what quantitative sociologists take to be the basis for the successes of the physical sciences. If we return to the simple formula, $F = ma$, only a very few variables have sufficed to yield extremely accurate predictions. Such formulas can help us sociologists to understand the importance of using concepts defined at very high levels of abstraction, of linking concepts with one another systematically—although not necessarily mathematically—and of testing theoretical ideas. Unfortunately, we have learned instead to isolate phenomena from their complex contexts so as to yield the kinds of measurements that prepare the way for using mathematics to help us make predictions.

It is important for us to examine the measurement procedures of the sociologist in order to understand more fully the fundamental problems associated with such quantitative approaches within sociology. Overall, efforts follow the direction of employing as much mathematics as possible, conforming to the interests of early philosophers of science who were steeped in mathematics. We have paid little attention to the pragmatist doctrine put forward by Abraham Kaplan:

> It is one of the themes of this book that the various sciences, taken together, are not colonies subject to the governance of logic, methodology, philosophy of science, or any other discipline whatever, but are, and of right ought to be, free and independent. (Kaplan 1964; quoted in Diesing 1991:82)

Instead of such independence, which would look to assess the actual achievements resulting from methodology, we have quantitative measurement procedures geared to the movement from nominal to ordinal to interval to ratio scales, supporting a focus on mathematical prediction. Further, the focus is on increasing the reliability and precision of measurements while ignoring the range of contextual factors involved. For example, there is interest in specifying the "operational definition" of a given concept, moving down language's ladder of abstraction.

Yet another aspect of efforts to move down that ladder has to do with procedures for obtaining probability samples, typically for surveys of a given population. Such procedures employ mathematical assumptions that enable the researcher to conclude, say, that a given sample of individuals represents a much larger population of individuals to within a specified degree of sampling error. And this in turn becomes the basis for quantitative analyses of the resulting data. However, the focus of such sampling procedures is generally quite narrow, once again serving to simplify enormously complex situations and enabling the researcher to move down the ladder abstraction and draw simplistic conclusions about the relationship between two or several variables. For example, it is a rare study that takes into account populations in the past as well as the present; the focus is on the present. Even the rare "panel study" takes place over a limited number of years, avoiding very long-term change like that from preindustrial to modern society or oral to literate society. If culture is indeed an important concept, then how are probability samples to tell us anything of significance about basic cultural change, granting the very rare sampling of written materials over long periods of time? Sampling procedures were originally developed not by sociologists but within other disciplines. For example, agricultural economists were interested in improving crop yields, geneticists wanted to learn about heredity, and the U.S. Armed Forces wanted to plan bombing runs during World War II (Schutte 1977:Appendix I). None of this speaks to our own problem of attempting to understand cultural change.

Integrating Knowledge

We turn now from sociology's movement down language's ladder of abstraction to its movement across the discipline's tower of Babel. Overall, what becomes obvious as we proceed with this analysis is the way in which all aspects of the sociologist's usage of the scientific method are intimately tied together within the same scientific paradigm. For example, tied closely to the above analysis of our orientation to problems and our movements up and down language's ladder of abstraction are our procedures for drawing statistical inferences about the relationship between two variables. Here once again we make substantial use of mathematics and attempt to move toward prediction. This results in statements with very limited utility either for understanding phenomena or for solving practical problems. For example, we may be able to reject the null hypothesis that there is no relationship whatsoever between, say, the number of people visiting a birth control center and the birth rate of a given people. In other words, we find that the birth control program "works" at least to a very limited extent, and this has at least some utility in efforts to evaluate its effectiveness. But what have we learned as a result? Theoretically, we have learned little more about the impact of cultural values and norms, of patterns of bureaucracy and social stratification, of anomie and alienation, of patterns of conformity and deviance, of relative deprivation and reinforcement. And there appears to be little practical impact for our conclusion, since we surely did not really previously believe the null hypothesis that the birth control program had absolutely no effect on the birth rate. And what have we learned about comparing this program with many others as well as how to improve any of these programs?

In an effort to answer at least some of these questions, quantitative analysts have utilized procedures for correlation and regression, both of which encourage precise measurement procedures. Efforts at correlation quantify the degree to which variation of a given variable is accounted for by variation in another variable, getting beyond the simple statement that there is at least some relationship between the two that could not easily arise as a result of sampling error or chance. And regression procedures specify mathematical formulas that we can use to make predictions from what we know about one variable to what we do not know about another. In our birth control example, correlation would tell us just how much the birth rate would be affected by a birth control program, and regression would enable us to make a prediction as to the change in birth rate from our knowledge of the existence of birth control programs. One problem, however, is that use of such more sophisticated quantitative procedures— which also require more assumptions that may not hold true—does not necessarily yield high correlations or accurate predictions. Arguably, sociological efforts to correlate and predict have generally yielded low correlations and inaccurate predictions. This is quite understandable once we take into account Blumer's critique of the analysis of variables as well as the general failure of such quantitative research to conceptualize variables at very high levels of abstraction. Rather than yield sophisticated knowledge, sophisticated tools can serve to show up our enormous ignorance.

One response on the part of quantitative sociologists to these problems is to move further in a quantitative direction. If high correlations and accurate predictions cannot be made from knowledge of one variable, how about many variables? For example, they use such procedures as factor analysis, cluster analysis, partial and multiple correlation, multiple regression, path analysis, and discriminant analysis. There are definitely occasional instances in which such procedures, when coupled with abstract theoretical concepts, have advanced our understanding, and one illustration will be presented in Chapter 4. And there are also instances when such analyses have yielded better bases for evaluation research, where higher correlations or more accurate predictions aided in the overall assessment of certain projects or procedures over others. Yet I believe it is arguable that such quantitative approaches have, in general, further diverted attention from abstract conceptualization and theory and taken us still further away from developing a sociology that is integrated, credible, and cumulates rapidly. Such a sociology requires us to face up to our present divorce between methods and theory. Quantitative procedures, by contrast, generally place mathematics once again in the saddle, riding the horse of a long-dead philosophy of science.

Efforts to integrate knowledge have not been limited to such relatively complex mathematical procedures. For example, we might have reference here to the cross-tabulational tradition within sociology, primarily based on procedures for the analysis of surveys. Historically, many of such procedures grew out of research sponsored by the U.S. Armed Forces during World War II, followed by great interest in survey research and accompanied by electromechanical inventions for the analysis of data with the aid of punched cards. "Cross-tabulation" is simply a way of obtaining the distribution of one variable within the categories of another. In a simple example, we might determine whether a greater percentage of those attending birth-control clinics had a lower birth rate than those not attending them. Sociologists have emphasized cross-tabulations, even to this day, in their search for cause-and-effect relationships between variables. Perhaps the greatest influence on procedures for examining such relationships was "the elaboration model," developed by Paul Lazarsfeld and his coworkers at Columbia University in order to interpret data obtained on the American soldier (Stouffer et al. 1949) during World War II (Lazarsfeld and Rosenberg 1955; see also Hyman, 1972; Phillips 1985:430–42). Such procedures often focus on secondary analyses, and they generally involve the introduction of a third variable and a detailed cross-tabulation of three variables. Many such analyses have yielded considerable insight, but generally they have contributed to the imbalance between theory and methods.

Worldviews

The fact that all of these problems associated with our present approach to the scientific method are intertwined with one another suggests not only the existence of a scientific paradigm or subculture that yields them but more generally a cultural paradigm within which that scientific paradigm is located. As for the nature of that cultural paradigm we can turn to several analyses, taking into account what sociology as a whole reveals about this matter. Let us begin with a different quote from Nietzsche, who appears to have captured much of the nature of our worldview:

On the doctrine of poisons—So many things have to come together for scientific thinking to originate; and all these necessary strengths had to be invented, practiced and cultivated separately. As long as they were still separate, however, they frequently had an altogether different effect than they do now that they are integrated into scientific thinking and hold each in check. Their effect was that of poisons; for example, that of the impulse to doubt, to negate, to wait, to collect, to dissolve. Many hecatombs of human beings were sacrificed before these impulses learned to comprehend their coexistence and to feel that they were all functions of one organizing force within one human being. And even now the time seems remote when artistic energies and the practical wisdom of life will join with scientific thinking to form a higher organic system in relation to which scholars, physicians, artists, and legislators—as we know them at present—would have to look like paltry relics of ancient times. ([1887] 1974:173)

For Nietzsche, "God is dead" implies the end of an entire way of life, granting that it will take some time for mankind to understand the nature of its new situation. As for the nature of our old way of life, Nietzsche's vision is that of a bureaucracy—granting that he did not employ that concept—where its various elements are unable to interact with one another. Nietzsche saw such lacks of integration as "poisons." And his vision of an alternative involved not just the unification of those poisons within an overall approach to the scientific method but also a unification of that scientific method with "artistic energies and the practical wisdom of life." Here, he saw science uniting with all other efforts to penetrate the mysteries of the universe, and here we can more fully understand his vision of "gay science" as anticipating postmodern critiques of Enlightenment science's emphasis on the rational at the expense of the emotional. Yet he viewed any such occurrence as taking place in the future: "And even now the time seems remote . . . " We contemporary sociologists need not, however, be dependent on Nietzsche's insights or metaphors to understand our situation, granting their usefulness up to a point. We can turn to our own concepts, such as bureaucracy, and our own analyses. Here we can go back to the above analysis of our own utilization of the scientific method. What does it tell us about our worldview?

For example, let us not ignore the deep aspirations within the discipline, even today despite growing cynicism and pessimism, for fulfilling Enlightenment cultural values. Those values—such as equality, freedom, democracy, science and secular rationality, and the ultimate worth of every single individual—appear to have grown far stronger since the eighteenth century, not only within sociology but also throughout contemporary society. Yet since we sociologists are not accustomed to taking culture seriously as a powerful structure in its own right, we generally pay little attention to those values. After all, this concept is the key conceptual tool of anthropology, whereas patterns of social organization like stratification, bureaucracy, and group point us toward a more unique view of sociology. And if indeed bureaucracy is a powerful force in our lives as research seems to demonstrate, then we have learned to divide up the labor as good bureaucrats should between anthropology and sociology. And in the bargain we need not concern ourselves with the nature of preindustrial society or oral society. We can still allow some sociologists to form a section on the sociology of culture, and we can still teach culture in our introductory courses, yet by and large we can continue to ignore that concept as a central one within our discipline. And we can continue to ignore the impact of that concept, revealed by studies in anthropology as well as sociology, for understanding our worldview.

Turning to key concepts revealing our patterns of social organization—social stratification, bureaucracy and group—when we link them to changes in cultural values we can begin to grasp the dynamic behind our worldview. For example, it appears that we have contradictions between, on the one hand, our egalitarian openness to knowledge, which the cultural value of science and secular rationality proclaims, and on the other hand, bureaucratic patterns of organizing our discipline, where studies of culture and social organization remain separated from one another. This contradiction seems to be played out within modern society as a whole, where aspirations for fulfilling those cultural values are limited by patterns of stratification and bureaucracy within our variety of groups. Further, if we take into account cultural change, where egalitarian and other humanistic values are emphasized more and more, what emerges is a growing gap between aspiration and fulfillment. For example, there is stratification within society as well as among ourselves of physical and biological scientists over social scientists, given the incredible technological achievements based on knowledge from the former. Never mind the complexity and dynamism of human behavior, never mind how the deemphasis on the importance of the social sciences has become a self-fulfilling prophecy, and never mind the enormous amount of knowledge that we social scientists have uncovered.

If our worldview is, then, structured by a bureaucratic ethos encouraging a growing gap between aspirations and fulfillment—as illustrated by the growing gap worldwide between the rich and the poor—we can also apply the concepts of anomie and alienation to our situation. Beyond Durkheim's dated view of anomie as normlessness—given our present understanding of the pervasiveness of cultural norms—we might well invoke a revision of his concept, as suggested by Merton (1949), to point toward our failure to fulfill basic cultural values even when we are acting in conformity to cultural norms. Here we would do well to include material along with nonmaterial cultural values, where anomie is illustrated by the failures of millions of workers to fulfill values of achievement and success, settling instead for temporary employment without health insurance, with the ever-present threat of unemployment and with rates of pay that make it very difficult to support a family. And given a growing gap between aspirations and fulfillment, we have increasing anomie within modern society. As for alienation, with our bureaucratic approach to sociology we generally relegate this concept to voting studies or Marxist sociology, just as we relegate anomie to studies of suicide or crime. Yet a more abstract view would tie these concepts to contradictions between cultural values and patterns of social organization within modern society, and also to an increasing gap between the two.

It might be argued that the implicit approach to sociology and the social sciences within these pages is overly optimistic, given our failures over the entire history of the discipline to develop a base of knowledge that might serve as a platform for cumulative technologies, by comparison with the physical and biological sciences. Haven't we already had enough time to test the Enlightenment dream of the classical sociologists, and isn't our present situation the best that we can expect? In my own view such questions are entirely legitimate, yet they fail to take into account the devastating impact of a bureaucratic worldview on sociology in comparison with the physical and biological sciences. It is one thing to divide up knowledge in fields where mathematical relationships work underneath the surface to integrate those fields and yield the basis for useful and accurate predictions employing very few variables. But it is quite another thing to chop up sociology and the social sciences where no such formulas are available. And our situation is worsened by our unbalanced emphasis on those mathematical tools instead of abstract concepts, illustrating our continuing allegiance to a long-dead philosophy of science. Granting the complex nature of human phenomena, such concepts can enable us to take into account that complexity, and they can be coupled with a systematic and broad approach that would parallel what the physical and biological sciences have achieved.

A SCIENTIFIC METHOD FOR SOCIOLOGY

Following Kuhn, it is the lack of a clear alternative to our present scientific and cultural paradigms that holds us back more than anything else, one where the problems and contradictions within those paradigms promise to be resolved. That is what I wish to sketch in this final section of the chapter. We shall proceed with five subsections describing elements of the scientific method covering topics that are much the same as those in the above five subsections criticizing our present approach to the scientific method. Following Nietzsche's analysis, the above five elements are "poisons" insofar as they are not integrated, kept apart by our bureaucratic worldview. Yet also following Nietzsche's ideas, combining these elements can work to place us on the same "firm ground" achieved within the physical and biological sciences over the past four centuries.

Definition of the Problem

Following the analysis in the above section, our worldview involves a fundamental contradiction between cultural values and patterns of social organization, a contradiction that appears to be increasing over time. No small aspect of our patterns of behavior is involved, but rather the two basic aspects of social structure. How are we to understand this state of affairs, granting that evidence for it will be presented in Chapters 3 and 4? If the scientific method calls for the definition of a significant problem, then this contradiction within the social structure of modern society passes the bar. This problem involves not just one small aspect of culture or of social organization, but a great deal of each. Further, we must look to the distant past, the present, and the future. Overall, given the breadth of this problem, if we are to develop an adequate definition of it then our usual specialized approach to the scientific method will prove to be ineffective.

One illustration of the increasing contradiction between cultural values and patterns of social organization derives from an international study of the change from preindustrial to industrial society (Lerner 1958:esp. 23–25). In the early spring of 1950 an interviewer named Tosun B. who lived in Turkey's capital city of Ankara journeyed several miles away to the village of Balgat. Since there was no road between Ankara and Balgat the trip took two hours by car. Tosun asked the village chief how satisfied he was with life. He replied:

> What could be asked more? God has brought me to this mature age without much pain, has given me sons and daughters, has put me at the head of my village, and has given me strength of brain and body at this age. Thanks be to Him.

The village grocer presented a different picture of his contentment:

> I have told you I want better things. I would have liked to have a bigger grocery shop in the city, have a nice house there, dress nice civilian clothes.

He had seen a movie portraying the kind of shop he wanted, with "round boxes, clean and all the same dressed, like soldiers in a great parade." But the grocer also sensed his limitations: "I am born a grocer and probably die that way. I have not the possibility in myself to get the things I want. They only bother me."

Tosun described the village chief, a sixty-three-year-old man, as "the absolute dictator of this little village." What would he do as president of Turkey? He would seek "help of money and seed for some of our farmers." The village grocer was "the only unfarming person and the only merchant in the village." According to Tosun "he is considered by the villagers even less than the least farmer." In their eyes he had rejected the worth of the community and even the supreme authority of Allah. His response to Tosun's question, by contrast with the chief, was not limited to what he would do for Balgat: "I would make roads for the villagers to come to towns to see the world and would not let them stay in their holes all their life."

Tosun's interviews with the chief and the grocer included this question: "If you could not live in Turkey where would you want to live?" The chief's response was "Nowhere. I was born here, grew old here, and hope God will permit me to die here." Only the grocer was able to imagine himself living outside Turkey, and he responded with an alternative: "America, because I have heard that it is a nice country and with possibilities to be rich even for the simplest persons."

If we employ the chief and the grocer as metaphors for stages—not necessarily inevitable—of the industrialization or modernization process, then they suggest the schematic diagram in Figure 1-1. The grocer's heightened material aspirations ("I would have liked to have a bigger grocery shop in the city, have a nice house there, dress nice civilian clothes") and nonmaterial aspirations ("I would make roads for the villagers to come to towns to see the world and would not let them stay in their holes all their life") illustrate the dramatic increase in the top curve or the revolution of rising expectations. By contrast, we have the chief's "What could be asked more?" his limited view of what he might do as president of Turkey (seek "help of money and seed for some of our farmers"), and his failure to conceive of ever living outside Turkey ("I was born here, grew old here, and hope God will permit me to die here"). The chief's level of aspiration locates him at the lower end of the revolution of rising expectations.

What Figure 1-1 suggests is that we take into account two curves simultaneously: the revolution of rising expectations, and the fulfillment of those expectations. One way to do this is to focus on the gap between the two curves. The chief's expectations appear to be quite limited, and generally they have been fulfilled, yielding a very small gap between the curves. He stated, "What could be asked more?" We might locate him on the left side of Figure 1-1 in the preindustrial period. By contrast, the grocer's aspirations are quite high and they remain largely unfulfilled, creating a large gap between the curves. Recall his statement, "I have told you I want better things." We might locate him within modern society on the right side

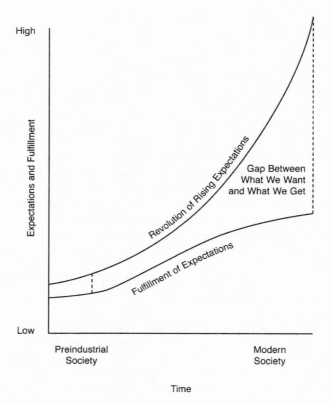

Figure 1-1. The invisible crisis: the escalating gap between expectations and fulfillment.

of Figure 1-1. To the extent that we are in fact dealing with the change from preindustrial to modern society, then Figure 1-1 suggests the existence of an exponentially deepening crisis, for the gap between the curves is growing exponentially. And to make matters far worse, it is the kind of crisis that is largely invisible and therefore will generally be ignored. Expectations are relatively intangible, especially when we are referring to widely shared ones. Further, it is the relationship between expectations and fulfillment that is at issue, and this relationship is more intangible than each curve taken separately. Still further, we are examining this gap over a very long historical period, making the idea in question quite invisible.

Historically, the top curve for the revolution of rising expectations refers to the dramatic cultural changes accompanying the shift from preindustrial to modern society, as illustrated by the grocer's "I have told you I want better things." We might look to the development of physical science and technology in the seventeenth century as fostering materialistic values. And we might also look to that development as the basis for eighteenth-century Enlightenment optimism about human possibilities. That Enlightenment era encouraged people-oriented values such as "liberty, equality, fraternity," and the nineteenth and twentieth centuries have expanded an emphasis on those values and extended them worldwide. If we shift from the top curve of Figure 1-1 to the bottom curve, which depicts our opportunities for fulfilling values, perhaps the central finding of sociology as a whole is the existence of social stratification—a species of social organization—in all aspects of modern life, whether it be in the form of classism, ethnocentrism, racism, sexism, ageism, or other isms that have received less attention. This pattern restricts people's ability to fulfill their cultural values, and it is not limited to restricting the opportunities of just some individuals. The values involved here are not merely material ones but people-oriented ones as well. And here, even billionaires suffer from forces limiting the fulfillment of those values, granting their far greater ability in the materialistic area.

High Level of Abstraction

What we have outlined above appears to be a fundamental and increasingly urgent problem within modern society. Our most powerful tools for addressing that problem are the abstract concepts of sociology, concepts that are linked together systematically rather than employed in isolation from one another. Yet to understand more clearly just how we can accomplish this and fly in the face of our bureaucratic approach to the scientific method, let us turn to an image that can yield insight into our ideals for the scientific method. Figure 1-2 can help us to understand language's levels of abstraction as well as links among various areas of specialization. This metaphor is partly based on an idea advanced by Lev Vygotsky ([1934] 1965:112), a social psychologist whose work included a deep interest in how language works. As we move down lines of longitude reaching from the North Pole to the equator within the northern hemisphere, we are also moving down language's ladder of abstraction from concepts that are very abstract to concepts that are very concrete. This movement works in the same way for the physical and biological sciences and the social sciences. Similarly, the South Pole represents a high level of abstraction for literary concepts, and movement up to the equator is movement to concepts at a low level of abstraction. However, no actual phenomena are depicted at the equator, but only concepts at a very low level of linguistic abstraction.

For example, within sociology we might move from "social stratification"—defined broadly enough to encompass racism, classism, ageism, sexism, and ethnocentrism—to the concept of racism between blacks and

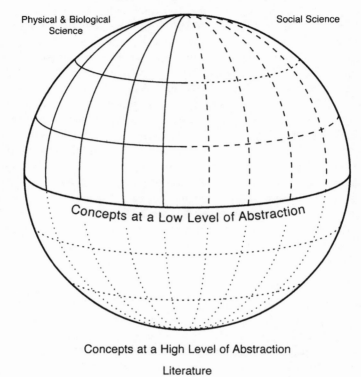

Concepts at a High Level of Abstraction

Literature

Figure 1-2. The globe metaphor for the languages of science and literature.

whites in American society, and still further down, black-white racism within American society after the Civil War. Within physics we might have concepts like "force" at the North Pole—defined broadly enough to include the force of gravity along with mechanical and other forces—and the force of gravity of objects in free fall near the earth still further down. Literature, by contrast, does not emphasize technical terms but communicates with ordinary language. And ordinary language is also organized by levels of abstraction, such as "human being" at the South Pole, "men" closer to the equator, and "Hamlet" at the equator. And literature with its metaphors can illustrate the power of language to jump from the concrete to the abstract, where "Hamlet" can come to mean the indecisiveness of all human beings when we are faced with problems demanding action. Metaphorically, we can all come to see ourselves as Hamlets to a degree.

From the perspective of the globe metaphor, all language works in similar ways, where its utility derives in large measure from its helping us to move from the general or abstract to the specific or concrete, bringing general experience to bear on any specific situation or problem.

The globe metaphor provides us with a clearer picture of what Willer and Webster attempted to teach us over three decades ago: it is concepts at a very high level of abstraction that we sociologists desperately need if we wish to learn from the achievements of the physical and biological sciences. Figure 1-2 does not focus on the importance of prediction or the use of mathematics as a direction for sociology, procedures that we sociologists have come to see as vital by accepting an outdated philosophy of science. Rather, it depicts movement far up language's ladder of abstraction, just as physics and biology as well as literature do so. That movement creates concepts that can function as broad umbrellas that enable us to then move very far down that ladder of abstraction to concrete experience. We see this in the physical concept "force," where we can come down to invoke any concrete force whatsoever, whether it be a falling apple or the motion of our galaxy. And we also see it in the use of "Hamlet" as a metaphor for the indecisiveness of all of us human beings, and we can then come down to apply that metaphor to any one of us. Here, then, is a clear direction for sociology and the social sciences, one that sharply contradicts Merton's emphasis on theories of the middle range: to conceptualize at very high levels of abstraction and use those concepts to carry much of the weight of what we have learned through our research over the years.

Figure 1-3 presents twenty-six key concepts generally taught in introductory sociology, studied in graduate school, and given at least lip service in sociological research. The top row above the dashes portrays social structures, with middle row depicting situations and the bottom row showing individual structures. These twenty-six concepts have been selected because of their importance within the discipline, their abstract conceptualization, their readiness to form links with one another, and their range of coverage. Many key concepts, such as ageism, authority, classism, collective behavior, community, crime, cultural change, demography, discrimination, ecology, economy, educational system, emotions, ethnic group, ethnocentrism, family, gender, industrial society, law, mental health, migration, occupations, political system, power, prejudice, race, racism, rational choice, religion, role, sex roles, sexism, social change, social class, social conflict, and social movements have been omitted. Yet the inclusive concepts in Figure 1-3 are meant to be supplemented by the range of less inclusive concepts within the discipline. For example, "institutions" would include "family," "educational system," "economy," "political system," and "religion." Also, "group" includes "community," "ethnic group," "race," and "social class." And "label" along with "social stratification" can

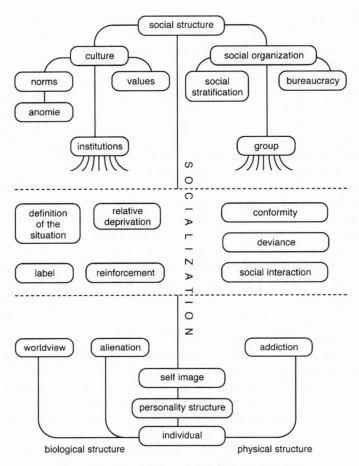

Figure 1-3. A web of sociological concepts.

be supplemented by "ageism," "classism," "discrimination," "ethnocentrism," "prejudice," "racism," and "sexism."

Many of the omitted concepts—along with still others that have been omitted but have not been listed above—are closely linked to those in Figure 1-3. For example, "authority," "power," "caste system," and "class consciousness" are linked to "social stratification" and "bureaucracy" along with the political and economic "institutions." "Ecology" is related to "physical structure" and "population" along with "race" and "gender" to "biological structure." "Emotions" are linked to "values," "relative deprivation," "reinforcement," and "alienation." "Law" and "crime" are tied to "norm," "value," "conformity," and "deviance." "Occupations" are related to "bureaucracy," "social stratification," and "group." "Role" and "sex roles" are linked to "norms" and "values" as well as "group" and "institutions." The approach to be adopted will emphasize change, taking up some of the slack resulting from the omission of "collective behavior," "cultural change," "migration," "social change," and "social movements." Other concepts were not added to this original list largely for the sake of brevity and simplicity. There is nothing holy about the list: to move beyond vague generalities about the importance of abstract concepts it is essential to choose some concepts and develop some illustrations.

A key to using Figure 1-3 as well as to using the web approach in general is an interest in applying simultaneously as many of these concepts as possible to any given situation or problem. For example, let us note the following three rows in Figure 1-3: social structure, the situation, and the individual. A wide-ranging approach to any given phenomenon would require that we dip into all three rather than divide up our scientific labor bureaucratically. In this way we combine a situational analysis with a structural one, and we also combine an interest in both social structure and structures associated with the individual. This kind of approach helps us to understand social and cultural change, for we are dealing with the momentary scene as well as long- and short-term structures. Further, we should also look to all three columns in our analysis: those headed by cultural "norms" and "values" and that headed by "social organization." We saw, for example, the importance of viewing both culture and social organization in our foregoing analysis of problems in modern society, where the two together enabled us to see a growing gap between aspirations and fulfillment. Metaphorically, these three columns have to do with the "head," the "heart," and the "hand," or the Wizard of Oz's Scarecrow, Tin Man, and Cowardly Lion. By dipping into all three columns we are responding to some postmodernist critiques of sociology in terms of its supposed overemphasis on rationality.

Several of the concepts in Figure 1-3 have not been emphasized within the contemporary sociological literature, yet they are useful for achieving wide coverage of phenomena. "Worldview" or *Weltanschauung* is a case in point, helping us to tie together the range of concepts in Figure 1-3, linking to the concepts of culture and cultural paradigm, and strengthening our conception of the nature and importance of the individual. Several earlier sociologists, such as Karl Mannheim (1952), saw the potential of this concept, but it takes an understanding of how it relates systematically to other concepts to realize that potential. "Biological structure" strengthens our understanding of the nature of the individual without any necessary tie to biosociology, and "physical structure" links to the importance of ecology as well as the physical sciences and the physical universe in general.

We might recall here that Karl Marx's approach to the idea of alienation, which typified the breadth of the classical sociologists, included physical and biological structures along with structures centered on society and the individual. As for "addiction," following its placement in Figure 1-3 it provides insight into patterns of individual action—similar to "habit" for psychology—which is missing from the sociological literature. Its abstract orientation goes beyond its link to physiological addiction, pointing to any outer-oriented compulsive behavior emphasized so much as to block out broader behavior, such as watching television, shopping, or running. We might note that these four concepts all are about individual structures, helping to repair sociology's one-sided focus on social structure.

It is important to understand that the concepts in Figure 1-3 are all seen as very abstract, bearing to a degree on all human phenomena rather than limited in time and space to certain eras or certain cultures, subcultures, or groups. With reference to the globe metaphor, they are all located at the North Pole. Many of us have been accustomed to seeing situational concepts—those located within the middle row of Figure 1-3—as being located further down those lines of longitude at a more concrete level. Perhaps this is because we can more easily experience momentary scenes than we can experience, for example, cultural values or bureaucracy. Yet those scenes cannot be understood without abstract conceptualization any more than large-scale changes throughout society. For example, W. I. Thomas employed visible examples in developing his concept of "definition of the situation," such as a mother's admonitions to her daughter to "Sit up straight," "Mind your mother," and "Be kind to sister" (1923:42). Yet on the basis of our present knowledge that definition of the situation also includes patterns of social stratification, cultural values, her own worldview, and much more besides.

We can see that stratification illustrated by the mother's commands to her daughter. We can also see indirectly that mother's adherence to cultural values emphasizing equality and individual worth in her saying, "Be kind to sister." And her outer-oriented worldview also comes through indirectly in her concern for her daughter's behavior and her effort to influence her daughter to look outward as well. Of course, there are a great many other ways of interpreting that mother's statements. What solid empirical basis is there for making these interpretations by contrast with other possibilities? What gives us the right to invoke patterns of social stratification, cultural values and worldviews on the basis of such limited knowledge of the mother's definition of the situation? Couldn't she have been invoking a much different definition? Of course, the scientific method guarantees no certainty nor even the promise of eventual certainty. But the above interpretation is based not merely on Thomas's quotes but on our entire web of sociological concepts and knowledge. We know a great deal about the patterns of social stratification and cultural values that were prevalent in the 1920s and are still widespread today. And we also are beginning to learn about the nature of our worldview then and now. We need not proceed from a bureaucratic approach to the scientific method with its focus on isolating each study and making precise predictions. Instead, we can adopt a more interactive approach.

This web of concepts suggests a more interactive scientific and cultural paradigm than our present bureaucratic ones. Scientifically, it points us away from our specialized tower of Babel and invokes the body of knowledge within the discipline as a whole. Both credibility and rapid cumulative development are at stake here. To the extent that we can indeed invoke that knowledge, then what we have to say will carry far more weight than the pronouncements of any specialized expert. Although it will not yield precise predictions, it will nevertheless give us the implications of many thousands of investigations and will yield much more profound understanding of the modern world than we presently employ. As for cumulative development, the language of sociology depicted in Figure 1-3 encompasses directly or indirectly current work in all sections of the discipline. It will not replace more specialized languages but rather interact with them and suggest both their further development as well as the further development of this general language of sociology. More concretely, it will open up paths for us sociologists to communicate with one another instead of hiding within our own special fields. To the extent that we buy into the importance of all three rows and all three columns of Figure 1-3 for understanding human behavior, then the work being done within every single one of our forty sections becomes relevant for every sociologist. As for our rationalization that we hardly have time to keep up with our own specialty, let alone other specialties, that becomes equivalent to the statement that we have no time to be sociologists.

Low Level of Abstraction

It is our ability to move down language's ladder of abstraction to concepts that are close in time and space to our concrete experiences—such as the colors and shapes that we see at this moment—that is so basic to the usefulness of ordinary language as well as literature. As for science, this very orientation is the basis for achieving the "firm ground" that Nietzsche saw as contrasting with the millennia of prescientific thought. Does our emphasis on highly abstract concepts somehow take away from the importance of moving far down language's ladder of abstraction? Within a bureaucratic worldview we tend to see movement up and movement down this ladder as working against one another. This is illustrated by the attitudes to philosophy as well as abstract theory held by many sociologists, following our tower-of-Babel perspective. Yet within an interactive worldview we can see such movements as swings of a pendulum, where the higher we move the further we can come down. Is this in fact possible? Does the web of abstract concepts described above help us to gain more insight into a given concrete phenomenon? The above illustration of Thomas's concept, "definition of the situation," also exemplifies such insight. When we see the mother's "Mind your mother" as illustrating social stratification, this alerts us to a great many other aspects of the situation that might otherwise escape us, such as her tone of voice, her volume, the rapidity of her speech, where she is located in relation to her daughter, her stance, her facial expression, and her daughter's facial expression and stance.

Using a highly abstract concept like social stratification, we can extend this approach to gain concrete insight into any phenomenon or situation whatsoever. And when we move from just one abstract concept to a web of twenty-six such concepts, we emerge with possibilities for gaining a great deal of insight into phenomena. For example, we might link Thomas's illustration to our own analysis of escalating problems within modern society. From that perspective we can begin to understand more fully the contradiction between the mother's cultural values of equality and the worth of the individual, exemplified by her "Be kind to sister," and her patterns of social stratification, illustrated by her "Sit up straight" and "Mind your mother." She appears to be reflecting, following our earlier analysis of this contradiction, a fundamental and escalating problem throughout modern society. That analysis invoked several abstract concepts in addition to that of social stratification: cultural values, bureaucracy, anomie, and alienation. And the other concepts among the twenty-six presented can flesh out this analysis much further, such as the worldview that keeps this growing contradiction in place as well as the mother's addiction to conforming to our outer-oriented and bureaucratic worldview. In this way we can move further down the ladder of abstraction for any phenomenon whatsoever, opening up to concrete details we would have otherwise ignored.

Yet our heritage from an earlier period when we looked to quantitative sociology as the basis for fulfilling the Enlightenment dream would argue against such a role for abstract theory. For example, one question that might be raised is that of obtaining agreement on specific operational definitions of any given concept. Other questions might have to do with the difficulty of obtaining valid, reliable, and precise measurements of abstract concepts. Such an orientation suggests a focus on prediction, which is certainly admirable as a long-term goal but which appears to distort our research efforts in the short run. We appear to require most immediately the ability to gain insight into complex problems with the aid of a system of abstract concepts that carries much of the weight of sociological knowledge. There exists an approach to measurement that points exactly in this direction. The idea of developing the "construct validity" of any given concept is an approach to measurement, developed within psychology (Cronbach and Meehl 1955), which points in this direction. Construct validity involves an assessment of just how well the empirical implications of a given abstract concept reflect what we have already learned from our web of abstract concepts. For example, does our concept of anomie suggest the same kinds of phenomena that are also suggested by our understanding of the growing contradiction between cultural values and patterns of social organization? Does it also reflect what we have learned about alienation and addiction?

Over the past several decades interest in ethnomethodology, rational choice theory, and symbolic interactionism has yielded a great deal of knowledge about concrete situations as well as procedures for learning about them. Mundane subjects such as conversations, greetings, arguments, and accounts of past experiences have been shown to reveal a world of complexity as well as patterns of behavior. Much of this knowledge has centered on small-group situations, as suggested by concepts in the middle row of Figure 1-3. Without such detailed situational knowledge we remain largely helpless in understanding social and cultural change. With it, we can combine structures with situations and penetrate the nature of change far more fully, yielding the potential for a genuine breakthrough in understanding human behavior. However, the problem with much of this work—in common with research within the rest of the discipline—is a failure to employ abstract concepts along with concrete ones. For example, there is a general avoidance of paying attention to culture and social organization, structures that we have learned a great deal about over the history of our discipline. To the extent that we come to see both situational and structural concepts as useful for analyzing any given scene, by contrast with our narrow bureaucratic interpretation of the scientific method, we will be able to probe ever more deeply into the complexities and dynamism within any given situation.

There is one area of concrete phenomena that we sociologists have tended to avoid, and we can understand this because of our outer-oriented worldview: past sociological research, as embodied in articles and books. Doing secondary analyses of such phenomena is similar to the hermeneutic tradition within philosophy, which originally centered on the interpretation of biblical and other sacred texts as well as on Roman law. Later, this approach was extended to any text whatsoever, with the definition of "text" extended to include any human action or product whatsoever. Some hermeneutic approaches have emphasized texts relevant to the social sciences (see, for example, Gadamer [1960] 1975; Apel 1980; Habermas 1971; Ricoeur 1970). What hermeneutics points us toward is the importance of language in concretizing any momentary scene, for that scene suggests the existence of an entire world. Yet we who interpret that text come from a different world, and what we have learned since that text was written can help us to understand both worlds. For example, we can look back at Marx's work and see, with the benefit of many decades of sociological research, the importance of culture and the situation in addition to patterns of social organization. And we can also see our failure to alter the worldview that deeply concerned Marx. And this "secondary" or hermeneutic analysis of Marx becomes a research study in its own right, one that is reflexive relative to sociological knowledge.

In addition to examining such classical research, we can perform secondary analyses of the full range of articles published in our journals along with our books. But doesn't this suggest concern with the number of angels dancing on the head of a pin rather than genuine sociological research? Shouldn't we be gathering new empirical data rather than rehashing old and dead material? Isn't such "research" no more than an exercise in navel-gazing? On the contrary, such research appears to be exactly what we need most in our situation of unconnected bits and pieces of knowledge and procedures that fail to follow the scientific ideals of openness to knowledge and the achievement of rapid cumulative development and a high degree of credibility. Secondary analyses can devote time and energy to conceptualization as well as the links among concepts rather than to the collection of still more data. They can point us in the very direction we have generally failed to follow in this century: toward movement far up language's ladder of abstraction. If we have already created a great imbalance favoring concepts at a low level of abstraction, then secondary analyses are urgently required to redress that imbalance. The fact that so few of them are published suggests the depth of that imbalance, for they generally emphasize abstract concepts more than primary analyses. Following this argument, they should be granted much higher priority than primary analyses at this time in history.

Integration of Knowledge

Just as our language consists not only of words but also of sentences and paragraphs, so does the language of science include hypotheses, propositions, and theories that link concepts with one another. Such statements make claims about the nature of the world, granting that they can never guarantee certainty. They can include existential or descriptive propositions, such as "Social stratification is widespread," as well as propositions pointing toward cause-effect relationships among concepts. And just as it is a web of concepts that is needed to assess the construct validity of any given concept, so is it a web of concepts within which any proposition about the nature of the world can be located. Within a bureaucratic approach to sociological research we do not look for that web, preferring instead to see propositions as isolated from one another. And we look toward using those propositions to make accurate predictions about phenomena. Further, we generally see those concepts as pointing in one direction only—from a given independent variable to a given dependent variable—yielding a relatively static approach to society. By contrast, our interactive orientation includes not only the impact of the independent variable but of the dependent variable as well. Also, our concern is with explanation far more than prediction. This is similar to our usage of concepts in ordinary language, where they yield insight without being located within propositions giving us precise predictions.

We can understand more clearly the nature of our web approach to the scientific method by looking to its philosophical origins in the work of Duhem (1954) and Quine and Ullian (1970) and their critique of approaches to positivism, as summarized by Kincaid (1996). Here, our globe metaphor can give us an image of that critique. Quine maintained that we cannot obtain concrete truths in isolation from abstract concepts. He saw the testing of propositions in a holistic manner, just as Mills saw the importance of shuttling up and down language's ladder of abstraction. Further, neither can we test propositions in isolation from one another, for it is an entire web of propositions that is tested simultaneously. Thus, any given concept or hypothesis should be seen as located within a "web of belief," with all parts of the web indirectly related to one another. Given this view of concepts and evidence, we sociologists have been interpreting the scientific method in a way that lost credence within philosophy long ago. Yet our general ignorance of this state of affairs is understandable, given the prevalence of a bureaucratic worldview. At this point, however, we can afford to be most optimistic. If our own interpretation of the scientific method has hurt us, then this is not any inherent barrier to understanding human behavior. Knowing this, we are now in a position to develop an interpretation closer to what we have learned about the way the physical and biological sciences have in fact worked.

We are now in a position, given our web of concepts, to return to the first step in our illustration of the scientific method, our definition of a problem. There we defined a mammoth and escalating problem within modern society: an accelerating gap between aspirations and fulfillment of those aspirations. What are the forces generating the long-term revolution of rising expectations as well as the fulfillment of those aspirations? Now, however, we have a web of concepts, and we can use some of them in a schematic diagram of forces behind those curves of aspiration and fulfillment. Here as before we shall present a schematic diagram meant to be suggestive. As for solid evidence, that is reserved for Chapters 3 and 4. Our approach will be to follow Kuhn's analysis of scientific revolutions, applying it to our sociological analysis of the problem depicted in Figure 1-1. We begin, using a pendulum metaphor, by pushing our pendulum as far as we can toward understanding the depth of that problem, taking into account social structures, individual structures, and situational forces. For example, given our accelerating gap between aspirations and fulfillment, just where does this appear to be leading us, considering in particular the possibilities of accelerating terrorism? And then in our final subsection on our worldview and a reflexive approach, we allow that pendulum to gather momentum in the opposite direction, taking into account the powerful forces involved in both our scientific paradigm and our cultural paradigm.

Figure 1-4 is based on an interactive approach versus a one-way approach to the relationships among phenomena (see, for example, Maruyama 1963; Forrester 1968, 1969; Meadows et al. 1972; Phillips and Senge 1972; Phillips 1980; Roberts et al. 1983; and Phillips 1985:84−95). The top loop centers on some social and individual structures and the bottom loop emphasizes several situational forces. Both are "positive loops," as conveyed by the circled plus signs in their centers, meaning that all of the forces involved continue to move in the same direction. The plus signs around the periphery indicate direct relationships. The top loop points toward society as a

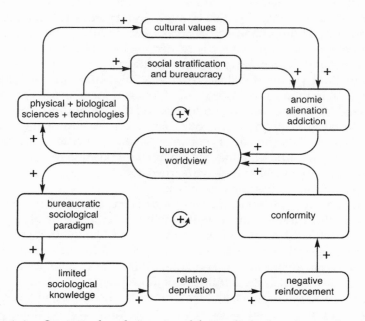

Figure 1-4. Structural and situational forces linked to the accelerating gap between aspirations and fulfillment.

whole, whereas the bottom loop centers on our situation within the discipline of sociology. Implicit in our joining the two loops is a conviction that what we do or fail to do within our own discipline has a marked effect on what happens within society as a whole.

The problem under analysis actually is far more complex than this sketch suggests. For example, forces opposing those depicted are in fact operating, preventing the continuing acceleration of each loop. Yet just as the simple schematic diagram in Figure 1-1 helped us to gain insight into the nature of this problem by omitting complexities, the simple schematic diagram sketched here can also be useful in helping us to understand the forces behind that problem by omitting complicating factors. There is a well-developed methodology that can take us from this relatively simple diagram to highly sophisticated procedures for computer simulation, as illustrated and explained in the above references to the work of Forrester, Roberts, and others. Such simulation procedures offer the possibility of carrying further the systematic approach to the scientific method adopted here. Specifically, they can deal with highly complex feedback relationships among a set of variables. However, the introduction of that methodology should await further commitment within our discipline to a scientific method that demands that level of sophistication. We already have too many examples within sociology of quantitative procedures that push us further and further away from a balance between abstract theory and data. Approaches to computer simulation that precede an overall understanding of a web approach to the scientific method can also serve to distract us in that way.

Beginning with the bureaucratic worldview in the center of Figure 1-4 and going around the top loop clockwise, that worldview gives impetus to the development of the physical and biological sciences and their technologies. Those technologies include procedures for constructing nuclear, chemical, and biological weapons of mass destruction. At the top of the loop we see the accelerating cultural values—for example, achievement and success along with equality and the worth of the individual—associated with our continuing industrial revolution, and these forces lie behind the revolution of rising expectations in Figure 1-1. Yet just below those cultural values are our patterns of social stratification and bureaucracy limiting the fulfillment of those values, as suggested by the bottom curve of Figure 1-1. Together, these forces of culture and social organization produce the widening gap depicted in that figure, a gap that is associated with anomie within social structure and alienation and addiction within the individual. These relatively invisible problems are linked to far more visible ones, such as a widening gap between the rich and the poor, patterns of discrimination, crime, terrorism, substance abuse, suicide, and divorce. Yet without an alternative worldview, these invisible and visible problems fail to yield serious questioning of our present worldview, and the latter continues to be reinforced as we go around the loop since it constitutes what we see as our only direction for solving those problems.

Proceeding counterclockwise from the bureaucratic worldview around the bottom loop, we come to our bureaucratic sociological paradigm and the limited sociological knowledge it produces, as we have argued above. We sociologists then come to feel relative deprivation, relative to the physical and biological sciences, for we are neither rapidly cumulating our knowledge nor developing a platform on which powerful social technologies can be constructed. The result is negative reinforcement, bearing in mind that our Enlightenment ideals still live and call for the development of our disciplinary knowledge in those ways. The resulting discouragement, cynicism, and pessimism in turn push us in the direction of hiding from this failure, yielding conformity to the norms within a given specialized field in our Tower of Babel. And such conformity in turn serves to reinforce our bureaucratic worldview, which, unfortunately, remains the only game in town. And in this way we continue to go around the bottom loop, all the while that anomie, alienation, and addiction continue to increase. Most of us continue to do our best in our efforts to develop knowledge and to apply that knowledge to social problems, and many of us raise questions about the existing structures of society or the existing procedures used within sociology. But our voices are easily drowned out in society, since we have gained little credibility and we speak more as individuals than with the voice of sociology.

Following Kuhn's argument and extrapolating it, we require not only a more interactive sociological paradigm but also a more interactive cultural paradigm or worldview as the broad framework within which that sociological paradigm is located. Figure 1-4 suggests that we sociologists can indeed have the impact on society called for by our Enlightenment ideals, provided that we are able to come up with alternative paradigms for sociology and society. That figure certainly does not offer us any detailed blueprint for how we might proceed, but its linking of situational and structural forces points us in a crucial direction. It is only in recent decades that we have come to stress the importance of situational factors, yet what we have failed to do is to link them with structural forces. When we proceed to do so we invoke the wide range of sociological concepts within the discipline. And that link appears to be fundamental to an understanding of social and cultural change. These feedback loops do not yield what most of us would like to have: an ability to make predictions with precision about the occurrence of a phenomenon or the solution of a problem. Yet an emphasis on exact prediction takes us away from gaining insight into how a number of structural and situational factors might come together to yield a given phenomenon. Pragmatically, this is much the way we approach phenomena and problems in our everyday lives: not with predictive formulas but with sensitizing ideas.

Reflexive Analysis and Interactive Worldview

Figure 1-5 sketches the potential impact of an interactive worldview and sociological paradigm on the problem depicted in Figures 1-1 and 1-4. The approach points toward continuing reduction of the gap between aspirations and fulfillments, as depicted in the graph at the top of Figure 1-5. The analysis of the forces involved parallels the analysis in Figure 1-4, where the basic change is the shift of the concept in the center of the two positive loops: from bureaucratic worldview to interactive worldview. Whereas the plus signs around the loops indicate direct relationships, the minus signs indicate inverse relationships. Moving counterclockwise from that concept to the bottom loop, it is that worldview that provides a framework for an interactive sociological paradigm, just as the bureaucratic worldview in Figure 1-4 encouraged the development of a bureaucratic sociological paradigm. Throughout the above materials I have argued that it is exactly this kind of scientific approach that is essential for the rapid cumulative development of sociological knowledge and the achievement of greatly enhanced scientific credibility. In turn, that achievement would yield greater recognition of the importance of sociology and less relative deprivation felt by sociologists in relation to physical and biological scientists. That in turn is a species of positive reinforcement, which should help to open up the sociologist to more social interaction with colleagues, given that the scientific method that works for us points to building bridges between specialists, and this in turn strengthens his or her interactive worldview.

Moving clockwise from that same concept of interactive worldview around the top positive loop, there is continuing impetus to the development of the physical and biological sciences and their technologies. However the development of highly credible sociological knowledge will increase awareness of threatening problems, and that awareness will play a role in decisions on the kinds of technologies that are developed. Cultural values like achievement and success as well as equality and the worth of the individual will continue to be emphasized, but the developing interactive worldview will point toward less stratification and bureaucracy throughout society. If this approach works well within our own discipline, there is reason to believe it would also work elsewhere. As a result, we might expect greater fulfillment of expectations, less anomie, less alienation, less addiction, and further support for and development of an interactive worldview. Although Figure 1-5 centers only on sociology among the various social sciences, there is also reason to believe that achievements within our own discipline based on a different interpretation of the scientific method will influence the other social sciences to follow our lead. This

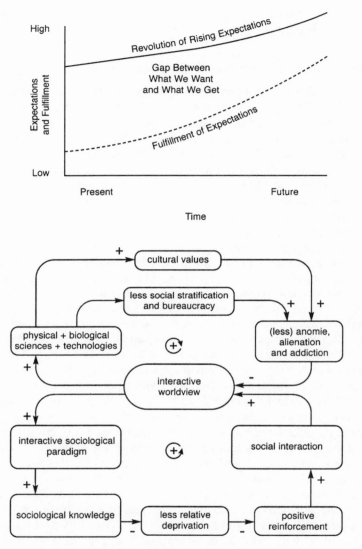

Figure 1-5. Structural and situational forces linked to the decelerating gap between aspirations and fulfillment.

need not involve adoption of our own concepts but rather what those concepts emphasize: the importance of taking into account a wide range of factors, such as social structure, the situation, the individual, culture, and social organization.

It is easy enough to draw optimistic diagrams, but is it in fact possible for sociologists to come up with the kind of powerful knowledge that in fact can affect forces as large and as invisible as anomie, alienation, and addiction within modern society? Even more difficult, can we hope to alter the very worldview and cultural paradigm that are basic to modern society? Following Figures 1-4 and 1-5, a key to understanding social and cultural change is our ability to take into account both structural and situational factors. To change our worldview, then, we require an alternative structure—another worldview—which promises to solve problems that could not be solved within the old worldview, such as anomie, alienation, and addiction. Further, from a situational perspective we must have a direction for how we can act in any given situation so as to follow the new worldview. Figure 1-5 illustrates this for work within sociology. However, there are all of the other situations that we all face in everyday life, and the new worldview must prove to be equally effective in all of those settings. More specifically, if we sociologists can learn to achieve positive reinforcement from developing sociological knowledge, we must learn to do the same from applying that knowledge in other situations. Here, Gouldner's call for a reflexive sociology and Mills's vision of the sociological imagination both point us exactly in this direction.

By contrast, our present bureaucratic ethos points us in an outer-oriented direction. We look outward, working to develop knowledge and communicate it to students as well as specialists attempting to solve problems within all of our institutions. Yet just as we have up to now failed to take a very hard and sustained look at our own interpretation of the scientific method, so have we also failed to examine our own bureaucratic role in society. Of course, many of us—such as C. Wright Mills—have criticized that role, but we have yet to see a direction for an interactive cultural paradigm that would alter it profoundly. If we began to adopt a more interactive approach to replace our outer orientation, then we would begin to look to our own usage of our abstract sociological concepts within our own everyday thoughts, feelings, and actions, following Gouldner's call for a reflexive sociology. And if we are to follow Mills's vision of the sociological imagination, then there is no reason why any individual could not follow along the same path. Such changes would not destroy existing bureaucratic organizations but would rather make them more and more interactive. And is it a utopian dream to expect that, if this occurred, all of our institutions would become increasingly effective in solving their basic problems, just as the technologies based on the physical and biological sciences continue to become more effective?

Our analysis in this chapter is limited in its communicating the exact nature of an interactive worldview as well as a reflexive approach to the scientific method and everyday life. Nevertheless, we have at least made a beginning in these directions. Given our deep involvement in a bureaucratic scientific and cultural paradigm, it is indeed difficult to speculate about an alternative worldview. Yet following the interactive relationships depicted in Figures 1-4 and 1-5 along with our pendulum metaphor for the scientific method, continuing efforts to uncover our fundamental problems can be expected to yield further understanding of an interactive and reflexive worldview. Following Gouldner, we cannot afford to restrict such efforts to sociological activities, for that restriction follows a bureaucratic ethos that separates work from life. And following our own analysis of the accelerating gap between aspirations and fulfillment, we must pay attention to our own structures of alienation and addiction, which have developed largely in response to that gap. More concretely, this translates into our learning to accept and deal with emotions like fear, guilt, shame, and anger. To the extent that we are successful there, we will be in a position to accompany our research with a reflexive analysis of our own impact on that research, and vice versa. And this would depart from our traditional bureaucratic and outer orientation, taking into account and learning about investigator effects.

Such a reflexive analysis should be seen as part and parcel of a scientific approach to sociology, for it follows scientific ideals of openness to all phenomena. If our justice system emphasizes the importance of self-interest in the acceptance of evidence and in the selection of jurors, how can we sociologists—who have learned so much about the impact of social interaction—ignore the self-interest or investigator effect of the social scientist as he or she proceeds to engage in social research? We might conceive of two kinds of secondary analyses or reflexive approaches, both of which fall within the hermeneutic tradition in philosophy and sociology, and both of which enable us to question the limitations of our worldview. One has to do with our taking a second look at some published material or some "text" defined far more broadly, and this involves an analysis from a different temporal and spatial perspective. This yields cumulative development in that we pay attention to a previous analysis, but we add to that analysis the understanding gained since it was completed. A second has to do with a secondary, hermeneutic, or reflexive analysis of one's own situation within the context of doing the research, yielding in a sense a double study: one on the external phenomena under examination, and one on the research situation itself. Both kinds of reflexive analysis are important within an interactive sociological paradigm and worldview, and both kinds are discouraged within our bureaucratic paradigms.

To illustrate briefly the beginnings of a reflexive analysis of the research situation, I might refer to my own hesitations, fears, and shame relative to the research problem of the accelerating gap between aspirations and fulfillment sketched in this chapter. Where do I get the nerve to point toward an alternative paradigm for the entire discipline of sociology? And far beyond this, who am I to propose nothing less than a change in our world-view if we are to achieve that alternative paradigm? Such hesitation, fear, and shame manifest themselves in many ways, such as burying my writing in endless qualifications, hiding behind the statements of a great many other sociologists and philosophers, writer's block, repetitive material, numerous drafts, and overly intellectualized writing.

Yet an understanding of the content of this illustration of the accelerating gap can help me to open up to these problems. For I am not alone in having them: if that gap is indeed wide and widening, then we all are under increasing pressure to avoid facing that gap if we have no way of reducing it. Or, to use sociological concepts, we have all become victims of anomie, alienation, and addiction. If I can recognize such problems even slightly within a given research situation—and if I have in mind alternative paradigms emphasizing that direction—then I can increasingly open up to such problems. I can even learn to transfer such behavior to the full range of problems that I experience in everyday life, giving me in turn a basis for learning to do reflexive analyses within sociology ever more effectively.

2

Cultural and Sociological Paradigms

The approach taken in Chapter 1 opens up two broad kinds of questions that relate most directly to the literature of philosophy and of sociology. First, there are questions about the nature of our cultural paradigm and worldview. These have to do with its nature, fundamental problems associated with it, whether we can conceive of alternatives, and whether those alternatives would retain the achievements of our previous cultural paradigm and worldview. These are very broad questions, and we can turn to philosophy for some ideas, just as we can turn to contemporary sociology for more specific ideas, such as the nature of our scientific paradigm. For example, we can learn what the achievements within sociology tell us about our scientific and cultural paradigms, and what they suggest with respect to alternative paradigms.

The first section of this chapter centers on questions relating to our cultural paradigm and worldview. Here we must go far back in time, even touching on biological evolution and the nature of the human being, for our worldview has been shaped by our entire history. John Dewey, writing his *Reconstruction in Philosophy* ([1920] 1948) just after World War I and his new introduction just after World War II, helps us to see those devastating events as stemming largely from our failure to achieve a reconstruction of our cultural paradigm and worldview and attempting to handle modern technology with a caveman mentality. Thomas Kuhn, a historian of science, presents in his *The Structure of Scientific Revolutions* (1962) a partial explanation of how fundamental changes in science have occurred and, by analogy, suggests the basis for changes in cultural paradigms. And Harold Kincaid's *Philosophical Foundations of the Social Sciences* (1996) gives us a very broad approach to the scientific method, one markedly different from what is presently in use within the social sciences yet highly appropriate for them.

Our second section centers on ideas about changes in sociology's scientific paradigm—following Kuhn's approach—which also takes into account the new cultural paradigm that Dewey called for and the approach to the scientific method suggested by Kincaid. Snow's review of sociology's achievements helps us to gain perspective not only on those achievements but also on our failures. And Scheff's treatment of "part-whole" analysis yields an approach that aims to fulfill Kincaid's vision of a web approach to theory within the social sciences. It is one thing to develop, as Dewey did, an overall orientation to philosophy and the philosophy of science coupled with a sense of urgency about the need for moving in that direction. Yet it is quite another thing to apply that orientation to the present situation of the social sciences, tackling the difficult methodological and theoretical problems involved. What is required, then, is nothing less than a reconstruction of the scientific method just as Dewey called for a reconstruction of philosophy.

CULTURAL PARADIGMS AND WORLDVIEWS

Although Dewey invokes the history of Western philosophy and thought in his efforts to come to grips with problems within our present cultural paradigm and worldview, it is essential that we go even further back in history if we are to gain the perspective we need for such an undertaking, even to the very nature of the physical universe. For we are biological creatures living in a certain kind of physical universe, and the biological and physical sciences have taught us—with a high degree of credibility—about our own nature and that of our environment. Here we might review the definitions of physical and biological structures that were advanced within Chapter 1 (see glossary). The glossary defines physical structure as a system of elements that interacts to a relatively small extent with its environment, by contrast with a biological structure, which interacts to a relatively great extent. Granting this difference, let us emphasize at this point what the two have in common: interaction with their environments. Ours is the kind of universe where no phenomenon can remain in complete isolation from any other phenomenon. For example, within our physical universe there can be no such thing as a perfect vacuum, for the walls of the container for a projected vacuum would be bombarded by phenomena from the rest of the universe and would transmit energy to the interior of the container, and those walls would also project outward the impact of phenomena from the inside.

What this suggests is that our cultural paradigm and worldview must somehow take into account these facts of life. Yet granting this similarity between physical and biological structures, it is also useful to examine the key difference between the two definitions: biological structures experience a relatively greater deal of interaction with their environments. For example, a tree's roots draw water and minerals from the soil, transferring them up to its leaves, and those leaves capture the sun's energy through photosynthesis and make that energy available to all parts of the tree. And the tree in turn continually affects its environment by reducing the amount of carbon dioxide and increasing the amount of oxygen. This momentary interaction between an organism and its environment, which is essential for the organism's very survival, is also illustrated by the long-term process of biological evolution. A species "learns" to adapt to its environments over long periods of time as a result of favorable mutations in an individual organism. Those mutations yield advantages for the organism's continued survival compared with other members of the species, mutations that are then inherited by the organism's offspring. For example, the long neck of the giraffe resulted from a mutation from shorter-necked giraffes, enabling the former to feed from the leaves of taller trees and giving offspring with long necks a decided advantage over their shorter-necked cousins.

This difference between physical and biological structures suggests a direction that our universe is taking, with no teleological implication that this direction has been foreordained or that it is inevitable: it is simply a matter of fact, and the facts supporting it have not been seriously questioned. The direction is from less to more interaction within the universe, and we can see this more clearly when we move to human beings with their personality structures and social structures built on the basis of language. From an evolutionary perspective, language gives the human being the capacity to speed up learning as well as adaptation to the environment by speeding up our capacity to interact with others, with our own past experiences, and with our physical and biological environments. Yet even this increased interaction has proven to be only a beginning for us. The change from oral to literate societies with the invention of writing and later of the printing press is a case in point. And with the invention of the scientific method and its enormous successes not only within the physical and biological sciences but also in the development of communication technologies like telephones, television, computers, cell phones, and the internet, our capacity to interact rapidly continues to increase. Given this situation of ever-increasing capacity for interaction and its implications for adaptation, why then should we humans be beset by urgent problems that threaten our very survival?

Our *capacity* for interaction does not necessarily translate into an *ability* to interact. I can easily communicate by e-mail with a great many individuals all over the world, but it remains an open question as to whether they genuinely take seriously what I have to say and look to the possibility of altering their own ideas and commitments on that basis. Similarly, I can receive a great deal of e-mail that questions my own ideas, yet how willing am I to pursue dialogues that may lead to abandoning my deepest commitments? All this has to do with the failures of the social sciences to learn how people can achieve genuine egalitarian interaction by contrast with our hierarchical patterns of social relationships, as illustrated in Figure 1-1. Following the top curve of our revolution of rising expectations, our aspirations increasingly call for the fulfillment of such cultural values as equality, which would establish a basis for egalitarian interaction. Yet the bottom curve of the fulfillment of aspirations appears to be dominated by such social structures as stratification and bureaucracy, creating an accelerating gap between aspirations and fulfillment. This would not prove to be so dangerous except for the fact that physical and biological technologies continue to yield ever more efficient methods of mass destruction. A little knowledge—knowledge from the physical and biological sciences but not from the social sciences—seems to be a dangerous thing, as illustrated in Figure 1-4.

From an evolutionary perspective, we humans appear to have the capacity—with language and the scientific method—to continue to interact and learn to adapt to an ever-increasing extent. Yet our partial use of that capacity within the physical and biological sciences—supported by a bureaucratic cultural paradigm and worldview—has created accelerating problems for our efforts to adapt to our environment. For we ourselves are rapidly changing our environment—and this includes ecological and social problems in addition to war and terrorism—so as to make it increasingly difficult for the human race to survive. The relevance of philosophy, and specifically philosophical efforts to examine the failures of our present cultural paradigm and worldview, becomes clear from this perspective. No meeting of heads of state, no action by the United Nations, no treaty for the nonproliferation of weapons of mass destruction, no election of any head of state, and no peace treaty compares with the importance of this task. The god of our cultural paradigm is supposedly dead, yet he somehow manages to live on in our hearts despite accelerating problems throughout the world that might give the lie to his importance.

John Dewey

If we look to Dewey's 1948 introduction to his *Reconstruction in Philosophy*, originally published in 1920, then what we have is a clear statement of his view of a key role of philosophy—and not just pragmatism—in the modern world: to analyze and understand the impact of science and scientific technology on human affairs. For the successes of the physical and biological sciences along with their associated technologies have, over the past four centuries, succeeded in transforming every aspect of our lives:

> The first step [of reconstructing philosophy], a prerequisite of further steps in the same general direction, will be to recognize that, factually speaking,

the present human scene, for good and evil, for harm and benefit alike, is what it is because, as has been said, of the entry into everyday and common (in the sense of ordinary and of shared) ways of living of what has its origin in *physical* inquiry. The methods and conclusions of "science" do not remain penned in within "science." Even those who conceive of science as if it were a self-enclosed, self-actuated independent and isolated entity cannot deny that it does not remain such in practical fact. It is a piece of theoretical animistic mythology to view it as an entity as do those who hold that it is *fons et origo* of present human woes. The science that has so far found its way deeply and widely into the actual affairs of human life is partial and incomplete science: competent in respect to physical, and now increasingly to physiological, conditions (as is seen in the recent developments in medicine and public sanitation), but *nonexistent* with respect to matters of supreme significance to man—those which are distinctively of, for, and by, man. No intelligent way of seeing and understanding the present estate of man will fail to note the extraordinary split in life occasioned by the radical incompatibility between operations that manifest and perpetuate the morals of a pre-scientific age and the operations of a scene which has suddenly, with immense acceleration and with thorough pervasiveness, been factually determined by a science which is still partial, incomplete, and of necessity one-sided in operation. ([1920] 1948:xxvii–xxix; emphasis added)

For Dewey, science and scientific technology must be at the forefront of this task of reconstructing philosophy along with our cultural paradigm and worldview. For science has succeeded in penetrating every corner of modern society and shaped our world. Yet if we look to Figure 1-4 we may note the negative role that the physical and biological sciences have played in providing the basis in knowledge for the production of weapons of mass destruction. Dewey in this passage gives recognition to science's role in the development of both "good and evil," "harm and benefit." And he recognizes the existence of opinions that see science as central to "present human woes." Yet it is not science per se which is at fault, but rather "a science which is still partial, incomplete, and of necessity one-sided in operation." It is physical and biological science without social science. He is quite explicit about our failure to develop the kind of science that can deal with "matters of supreme significance to man—those which are distinctively of, for, and by, man." He even goes so far as to claim that that kind of science is "nonexistent." For Dewey, then, science can save us, but what we require is not more effective physical or biological science but the development of effective social science. And the harm resulting from technologies based on the physical and biological sciences is not the fault of science but rather our failure to develop the social sciences.

Dewey looked to "Francis Bacon of the Elizabethan age" as "the great forerunner of the spirit of modern life," giving us hints we need today for reconstructing a philosophy or cultural paradigm that has yet to catch much of that spirit. Bacon found himself confronting a sixteenth-century orientation to worship the wisdom of the ancients. This backward look implied a negative view of the efforts of his contemporaries to achieve much in the way of learning and understanding. By contrast, Bacon condemned the inadequacy of existing knowledge, for he claimed that "knowledge is power," and he saw little of ancient knowledge that gave mankind much in the way of power to solve problems. He saw such knowledge as deriving from the worship of four kinds of idols. The Idols of the Tribe have to do with "the influence which the will and affections, such as pride, hope, impatience, and the like have upon the mind, but particularly the part which desire plays in determining thought" (Jones 1965:47). The Idols of the Cave refer to biases imposed on the individual by education specifically and social structure in general, yielding servility to the authority of the ancients. Idols of the Theater also refer to the individual's conformity to social structure, pointing directly to abstract philosophical systems that have little purchase on concrete experiences. And the Idols of the Marketplace invoke problems within language itself, for words are defective in representing things that don't exist or else confused ideas.

Bacon saw much that was negative even in his own day in the emphasis on specialization. A historian wrote this about Bacon in 1961:

> One of these impediments [to the advancement of learning], specialization in research, constitutes a virtue in modern eyes. The basis of his own method was a true, universal natural history, and he looked with suspicion upon any endeavor which sought to encompass only a small field. He finds fault with investigators' confining themselves to some one subject as an object of research, such as the magnet, the tides, and the heavens (he undoubtedly had Copernicus and Gilbert in mind), because he thinks it is unskillful to investigate the nature of anything in itself, since the same nature is also manifested in other things. He maintains the same attitude towarde the arts and professions, which, he insists, draw their strength from universal philosophy, and which, when studied individually, retard the progress of learning. (Jones 1965:46–47)

In looking to "the arts and professions" as well as the sciences, Bacon reaches outward from science so as to include technologies, seeing the industrial revolution as deriving from knowledge within the sciences and seeing developments within the humanities as linked closely to the new spirit of science. Specialization, for Bacon, generally involved turning away from a broad approach to the scientific method.

What Bacon called for is a new spirit of discovery, and not a continuation of the old emphasis on argumentation, proof, and persuasion, an approach where teaching meant indoctrination and the development of disciples. Such discovery for Bacon involves active experimentation, for scientific principles are not easily found on the surface of phenomena. Rather, they are hidden, for the world is complex. He elaborates on the differences between these different approaches to logic and the scientific method:

> Men have entered into the desire of learning and knowledge . . . seldom sincerely to give a true account of their gift of reason, to the benefit and use of men, but as if they sought in knowledge a couch whereon to rest a searching and wandering spirit; or a terrace for a wandering and variable mind to walk up and down with a fair prospect; or a tower for a proud mind to raise itself upon; or a fort or commanding ground for strife and contention; or a shop for profit and sale; and not a rich storehouse for the glory of the creator and the relief of man's estate. (quoted in Dewey [1920] 1948:57–58)

If biological evolution has given the human being the *capacity* to learn from experiences, excelling all other creatures in this regard, then the scientific method gives the human being the *ability* to learn—and learn very rapidly—from experience, an ability ignored by those who fail to understand the nature of this method.

Dewey viewed this discovery-oriented approach to the scientific method as the basis not just for changes in philosophy or in the social sciences but as a foundation for modern society as a whole. For example, in the area of morality he castigated the traditional focus within ethical theory on final ends or ultimate laws. Instead, he argues that we replace a focus on fixed ends such as health and amiability with a focus on progress in such phenomena, which also takes into account the detailed context of any specific situation:

> *Moral* goods and ends exist only when something has to be done. The fact that something has to be done proves that there are deficiencies, evils in the existent situation. This ill is just the specific ill that it is. It never is an exact duplicate of anything else. Consequently the good of the situation has to be discovered, projected and attained on the basis of the exact defect and trouble to be rectified. It cannot intelligently be injected into the situation from without. Yet it is the part of wisdom to compare different cases, to gather together the ills from which humanity suffers, and to generalize the corresponding goods into classes. Health, wealth, industry, temperance, amiability, courtesy, earning, esthetic capacity, initiative, courage, patience, enterprise, thoroughness and a multitude of other generalized ends are acknowledged as goods. But the *value* of this systematization is intellectual or analytic. Classifications *suggest* possible traits to be on the lookout for in studying a particular case; they suggest methods of action to be tried in removing the inferred causes of ill. They are tools of insight; their value is in promoting an individualized response in this individual situation. (Dewey [1920] 1948:169)

Within contemporary sociology many of us are accustomed to viewing any discussion of morality as out of bounds, granting that very recently a number of sociologists have begun direct exploration of this area. Yet if we return to Figure 1-3 and look to the stance of including more and more of that wide range of concepts in analyzing any given situation, then what emerges is nothing less than what Dewey prescribes. For example, Dewey's advice that we "generalize the corresponding goods into classes" is advice to use not just the concept of cultural values but also to develop more specific concepts detailing specific cultural values, like equality, achievement and success, freedom, and material comfort. His major argument, however, implies that we go beyond a focus on such values so as to include the full range of situational concepts specified within Figure 1-3. In this way the search for moral solutions to any particular problem becomes a contextual one that is solved—to a degree and only tentatively—by research. By standing outside the specific situation and pushing for the morality of this solution or that one, we fail to take into account much of what we already know about the impact of our actions, knowledge embodied within the range of concepts depicted within Figure 1-3. And as a result, we lose touch with the concrete consequences of our actions for the entire range of our cultural values, consequences that could be revealed by that knowledge.

This is not to suggest that the application of social science research to moral questions will solve those questions to the satisfaction of all concerned, granting it will help to clarify a great deal of what is at stake within any projected solution. Dewey's argument implies far more than the utilization of procedures such as those outlined above, procedures that he was not aware of in his day and that we are only now beginning to explore. He is calling for nothing less than a change in our cultural paradigm and worldview, something that will alter every single one of our institutions and the ways of thinking, feeling, and acting of every single individual. Metaphorically, we might see what he points toward as movement from a seesaw to a stairway, and we might apply these metaphors to current debates over a woman's right to have an abortion or, alternatively, a fetus's right to life. On a seesaw, with its emphasis on social stratification as well as conflicting cultural values—like freedom and equality as applied to the mother versus the fetus—there will be no resolution to the conflict and it might even escalate. On a stairway with very wide steps, by contrast, one person's gain need not be another's loss. For example, we might employ research on the genesis of forces such as anomie and alienation, which contribute to the existence of unwanted children, and with such knowledge we might reduce the demand for abortions, with both sides in the controversy moving up the stairway.

Of course, such vague metaphors are no more than a beginning in understanding the nature of our cultural paradigm, an alternative one, and how we might in fact move from the former to the latter. Dewey himself helps us a bit more when he singles out education for specific comments, contrasting education as merely a means to subsequent ends with education as a continuing process that is of value in its own right:

> Education has been traditionally thought of as preparation: as learning, acquiring certain things because they will later be useful. The end is remote, and education is getting ready, is a preliminary to something more important to happen later on. Childhood is only preparation for adult life, and adult life for another life. Always the future, not the present. . . . If at whatever period we choose to take a person, he is still in process of growth, then education is not, save as a by-product, a preparation for something coming later. Getting from the present the degree and kind of growth there is in it is education . . .
>
> We exaggerate the intellectual dependence of childhood so that children are too much kept in leading strings, and then we exaggerate the independence of adult life from intimacy of contacts and communication with others. When the identity of the moral process with the processes of specific growth is realized, the more conscious and formal education of childhood will be seen to be the most economical and efficient means of social advance and reorganization, and it will also be evident that the test of all the institutions of adult life is their effect in furthering continued education. Government, business, art, religion, all social institutions have a meaning, a purpose. That purpose is to set free and to develop the capacities of human individuals without respect to race, sex, class or economic status. And this is all one with saying that the test of their value is the extent to which they educate every individual into the full stature of his possibility. Democracy has many meanings, but if it has a moral meaning, it is found in resolving that the supreme test of all political institutions and industrial arrangements shall be the contribution they make to the all-around growth of every member of society. (Dewey [1920] 1948:183–86)

Dewey's prescription for us is to adopt a cultural paradigm and worldview based on continuing education as well as employment of the scientific method throughout society. We all should shift from a seesaw to a stairway paradigm throughout our lives, with the scientific method as the chief tool for helping us to move up the stairway. His continuing faith in education and the scientific method is indeed awe-inspiring, given that he wrote his book immediately after World War I and his new introduction shortly after World War II, and given that he did not have specific ideas on just how the social sciences might be employed to help attain the ends he envisioned. From his perspective, then, education at every stage from birth to death should be given fundamental value by the individual and society and should open up possibilities for future education. The economic institution should not merely produce and distribute wealth: it too has a fundamental responsibility to educate every individual to see the acquisition of wealth within the context of the widest possible range of cultural values and to move toward their fulfillment. The political institution, in addition to distributing power democratically, should take responsibility for empowering every individual to learn how to solve the full range of problems one confronts in life. The religious institution should teach every individual how to derive fundamental meaning from experiences—whether catastrophic or mundane—in everyday life. And so on with all other institutions.

This focus on some of the ideas of John Dewey should not imply any lack of importance of the long history of ideas in general and philosophy in particular. Neither should it suggest any lack of importance of the work of other pragmatists, such as Charles Peirce, George Herbert Mead, and William James. What Dewey gives us is a deep commitment and a sense of urgency relative to the task of changing our cultural paradigm and worldview, just as Nietzsche achieved with his pronouncement, "God is dead," some four decades earlier. Dewey also gives us, beyond Nietzsche, some systematic ideas about the failures of our present worldview as well as the approach we might take to the scientific method along with the kinds of institutions we might create. For example, Dewey helps to make Figure 1-2, with its contrast between the physical and biological sciences and the social sciences, come alive. He catches us sociologists up short by suggesting that our own failures are linked to the fundamental failures of modern society, and vice versa. We can also view Dewey's ideas from the perspective of what we know about biological evolution, that slow "learning process" that has given us humans the capacity for a very rapid learning process. The physical and biological sciences have translated that capacity into an ability, but this one-sided development threatens us all. What we require, following Dewey, is an ability to learn—based on the scientific method—that is taught within all of our institutions and that alters every individual's ideas, feelings, and actions from one moment to the next.

Thomas Kuhn

In Chapter 1 we touched on Kuhn, a historian of science rather than a philosopher, whose analysis of the process of change within the physical and biological sciences has proved to be extraordinarily useful and influential within the philosophy of science as well as throughout the social sciences (Kuhn 1962, 1977, 1992). Kuhn centered not on the ideals of the scientific method but on actual practices, and he found them most wanting in relation to those ideals, looking to such factors as tradition, hierarchy, and the orientations of the individual scientist as conflicting with ideals for how the scientist should in fact behave. He saw those factors as standing in the way of scientific revolutions, and he also saw them as tied to the prevailing scientific "paradigm" or subculture within the culture of society. In addressing the question of what it takes to actually achieve a scientific revolution, he helps us to understand the more general processes of social and cultural change, such as changes in our cultural paradigm. Of course, we must take into account Kuhn's immersion not in the social but in the physical sciences. For example, Dewey by contrast with Kuhn gives us the beginnings of a cultural paradigm pointing up the centrality of individual development within every single one of our institutions. Yet Kuhn's idea of paradigms derives from a systematic orientation involving hidden assumptions. And Kuhn helps us to understand the difficulties and also the possibilities of achieving fundamental changes in paradigms.

What Kuhn achieved in his extremely influential analysis of revolutions within the physical and biological sciences (1962, 1977) was nothing short of an intuitive application of the sociology of knowledge to explain dramatic achievements or revolutions within such sciences. It is not evidence alone that sways the community of scientists to reject an old theory in favor of a new one. Other factors as well play key roles, such as tradition, social hierarchy, and the personality of the scientist. Those factors parallel sociological concepts such as culture, social stratification, and personality structure. Scientific knowledge, then, emerges from some given social and historical context, and it must be understood in relation to that context, as opposed to some holy or unquestioned set of truths. Kuhn's most general way of understanding the forces operating within the work of the scientist is to see them as tied to the prevailing scientific paradigm. This concept, which variously suggests a basic example, a worldview, a model, profound beliefs, or fundamental assumptions, has been criticized by many academicians for its vagueness, and Kuhn himself abandoned it (1992:19). Yet at the same time it has been and still is widely used. If we see "paradigm" as an example of a "subculture," we can come to understand its vagueness as resulting from the breadth of this latter term and see it more clearly in relation to the concept of culture and, more generally, in relation to the literature of the social sciences.

Kuhn's approach is illustrated by his analysis of Einstein's development of the special theory of relativity. The American engineers Michelson and Morley used an ingenious series of mirrors located miles apart to measure the speed of light as it ricocheted off one mirror after another. Newton's laws apply to the motion of light no less than to the motion of a boat on a river: the boat is slowed when it moves upriver against the current or across the river, and it is speeded up downriver. Similarly, following Newton, light should be speeded up in the direction of the earth's rotation and slowed as it proceeds in other directions. Not so, according to the precise experiments of Michelson and Morley. Contrary to Newton's laws of mechanics, which had been widely accepted for hundreds of years, Michelson and Morley discovered that light travels at the same speed in any direction. Yet the physicists of Einstein's day, having lived their entire lives taking for granted the adequacy of Newton's laws of motion, generally remained unable to question those laws on the basis of experiments that challenged them. They were unable to follow the ideals of the scientific method and give appropriate weight to evidence against those laws. Einstein's theory not only had to confront specific evidence about the speed of light but also subcultural beliefs, stratification among scientists, and deep personal commitments.

Einstein somehow was able to challenge the prevailing paradigm or subculture of physical science by conceiving of an alternative paradigm, one based on a different way of thinking about motion (for a recent account, see Holton 1996). He started with the assumption that light travels at the same speed in any direction, thus accepting the conclusions of Michelson and Morley as a premise. This led to a number of strange hypotheses, which were implied by his special theory of relativity. For example, light should bend due to the force of gravitation. Also, individuals moving away from the earth near the speed of light would have their aging processes largely suspended, returning long after everyone they knew had died. We might see Einstein's view of the problem within Newtonian theory posed by Michelson and Morley as a first step of the scientific method, with his own theory, which he developed in an effort to solve that problem as a second step. A third step involved the empirical testing of Einstein's theory, and it has indeed been validated. That theory not only resolves Newtonian contradictions but is able to make more accurate predictions of motion in all situations, granting that Newtonian laws can still be used as an accurate means of making predictions in almost all situations. Following Kuhn's analysis, however, other factors were involved in the relatively slow process that ultimately led to the acceptance by physicists of Einstein's special theory.

Kuhn's analysis suggests that taking the first step of the scientific method requires some vision of subsequent steps of that method. Einstein, for example, could challenge Newton's laws because he was able to conceive of an alternative theory that could account for observed contradictions. By contrast, other physicists had no such alternative on which to fall back on and as a result were unable to take that first step. However much they might have desired to follow scientific ideals, they were unable to do so without that alternative. We might think of the scientific method as a kind of pendulum. We cannot confront our problems (step 1) without some way out once we face those problems (steps 2 and 3). Here, step 1 corresponds to a swing in one direction, and steps 2 and 3 to a swing in the other direction. A deepening awareness of problems (step 1) can push us to think about solutions (step 2), as well as testing those solutions empirically (step 3). That in turn can enable us to face up to our problems more fully, giving us the momentum to swing our pendulum in the opposite direction (step 1).

Carrying Kuhn's argument further than he himself did, postmodernists such as Latour (1987) and Rorty (1987) have opened up questions about the nature of the scientific method. Is the scientific method no more than a social process of negotiation where actors look to build unassailable networks of supporters? Is it nothing more than one kind of dialogue, which should not be privileged more than any other dialogue? Is the idea of "the scientific method" a bankrupt idea and scientific credibility no longer a desirable goal? Kuhn himself rejects such arguments:

> Interest, politics, power, and authority undoubtedly do play a significant role in scientific life and its development. But the focus taken by studies of "negotiation" has, as I've indicated, made it hard to see what else may play a role as well. Indeed, the most extreme form of the movement, called by its proponents "the strong program," has been widely understood as claiming that power and interest are all there are. Nature itself, whatever that may be, has seemed to have no part in the development of beliefs about it. Talk of evidence, of the rationality of claims drawn from it, and of the truth or probability of those claims has been seen as simply the rhetoric behind which the victorious party cloaks its power. (Kuhn 1992:9)

The questions about the validity of the scientific method that have been raised by many postmodernists are important ones, and it is most useful at this time to rethink the nature of that method and to learn from this controversy. Such critics of the scientific method have succeeded in shattering the claim that the findings of scientific research guarantee correspondence to truth or to the ultimate nature of reality, just as David Hume taught us many years ago that we can never be absolutely certain of the existence of any cause-and-effect relationship. Further, we do not even have any guarantee that this method will yield ever deeper understanding. Perhaps most important, such critics influence us to reexamine our approach to the scientific method, a direction that is most useful for the social sciences. Kuhn's reaction to attacks on the scientific method was not to reject them completely, for he himself argued that the scientific method must be understood not as yielding God-given truth but rather as a human effort that must be understood within its social and historical context. Granting all of the limitations of the scientific method, Kuhn argues that it nevertheless remains our best hope of achieving ever greater understanding of the world. Provided that we give up on the idea of certainty, or even that this method necessarily yields the cumulative development of knowledge, no other procedure offers us similar prospects for achieving cumulative understanding.

One implication of Kuhn's work for sociology is just this optimism about the importance of the scientific method, despite both postmodernist critiques as well as the blame heaped on that method for the present state of the world. He shares with Dewey a faith in the importance of science for modern society. At this time in history, when sociologists are struggling in their efforts to create a science that is both cumulative and credible and when there are serious doubts about this possibility, such optimism is in my view desperately needed. Another implication, as suggested by the concept of paradigm, is that we might do well to look to and make use of our own concepts of culture and subculture. For Kuhn is reinforcing what we have already learned within thousands of contexts: cultures and subcultures are powerful structures. A third implication derives from the systematic nature of the concept of paradigm with its view of ideas as resting on assumptions that are often hidden. Such an approach is useful in our efforts to understand subcultural and cultural change. For we can work to make visible what previously was invisible, just as Kuhn made visible patterns like social hierarchies, which worked to preserve allegiance to Newtonian theory. Such visibility can then alert us to contradictions between what was uncovered and scientific ideals. Following Kuhn's implications, we sociologists require an alternative subcultural and cultural paradigm if we wish to alter our existing ones.

Kuhn's work helps us to understand how the scientific method actually works. Given the importance of the scientific method for the changes in modern society, he conveys ideas about how we might proceed to change not only our subcultural but also our cultural paradigm. This involves changing fundamental structures in society, including all of the structures depicted in Figure 1-3: culture, social organization, and the individual. Those structures along with the other structures and situational factors in that figure are all linked together, with change in any one—such as culture or worldview—involving changes in all of the rest. The complexity of this can help us to understand why Kuhn claimed that it is very difficult to change a scientific paradigm. Here, we can add that it is far more difficult to change a cultural paradigm. Nevertheless, despite this difficulty, Kuhn gives us a direction for such change. For example, we sociologists are confronted with a huge contradiction between our Enlightenment ideals for our discipline and our present performance. And we might also add the contradiction between our scientific ideals for open communication coupled with our tower of Babel. Kuhn implies that instead of burying those contradictions we learn to make them visible, to construct an alternative paradigm for our discipline—and, we might add, for the culture of modern society—and then to test the utility of those paradigms for resolving those contradictions.

Yet given this Kuhnian approach to achieving fundamental changes in society, we still have a long way to go in learning how to achieve them. Among all of the academic disciplines it is philosophy more than any other that, historically, has come closest to probing in a systematic way the nature of our cultural paradigm or worldview, as illustrated by work in such fields as metaphysics and epistemology. In the following major section of this chapter we will take up sociology's own contribution to our understanding in this area, but in this next subsection we will continue with what philosophy has taught us. John Dewey, for example, has given us a general contrast between an unreconstructed and a reconstructed philosophy, with the latter calling for changes in all of our institutions as well as our worldview. Those changes require a focus on the individual's continued learning and development throughout life, something to which we give lip service and that our cultural values call for. Following Dewey's implications, we lack the fulfillment of those values largely because of the failures of the social sciences as well as our lack of understanding of how the scientific method actually works. Kuhn's analysis helps us with that understanding, yet we can proceed much further with it based in part on what has been achieved within philosophy since Dewey's time. Let us then return to the discipline of philosophy with our eyes on how we might learn to use the scientific method far more effectively within sociology.

Harold Kincaid

Modern philosophy like modern sociology is highly specialized, yet let us view it from the perspective of the broad approach of Dewey and the paradigmatic orientation of Kuhn. Both men set the stage for our understanding of the enormous impact of science and technology on our world, and both men saw fundamental problems in the pursuit of the scientific method. For Dewey it is the one-sidedness of a focus on the physical and biological sciences along with their technologies with a corresponding neglect of the social sciences that is basic to modern problems. Since science so thoroughly shapes our world, then a broader approach to it will have repercussions on all institutions and individuals. By reconstructing philosophy to accomplish this, we also prepare the way for basic changes in society and the individual. Kuhn centered on the scientific method in general and scientific revolutions in particularly, yet when we draw analogies to society as a whole we tread on the same ground over which Dewey walked. His paradigmatic orientation to scientific revolutions focuses on the process of changing subcultures and, by analogy, cultures. However, neither Kuhn nor Dewey treats in any depth the nature of the scientific method, especially as it might be applied to the social sciences. This is where modern philosophy can help the sociologist, giving us more specific ideas as to just where we are failing and where we are succeeding in our approach to the scientific method.

Harold Kincaid's *Philosophical Foundations of the Social Sciences* (1996) helps us to understand the enormous gap between a long-dead philosophy of science, which is still the basic orientation within sociology, and current philosophy of science. In a nutshell, the former—which we may label as logical empiricism or positivism—is bureaucratic whereas the latter is most interactive. Logical empiricists like Carnap, Popper, and Reichenbach generally had a background in the physical sciences and mathematics, viewing science as requiring a predictive and mathematical approach. Further, their focus generally was on hypotheses in isolation from one another. Social scientists have generally accepted this approach even while it was rapidly losing ground among philosophers of science. For example, social scientists can look to isolated hypotheses, or they can look to locating any given hypothesis within a web of hypotheses. Both Dewey and Kuhn were oriented in the latter direction, and Kincaid makes this orientation more explicit:

> Following Duhem (1954), Quine (Quine and Ullian 1970) argued that hypotheses do not confront experience or evidence one by one. Rather, testing a single hypothesis requires a host of background theory about the experimental apparatus, measurement theory, what data are relevant, what must be controlled for, and so on. So, when experiments fail, they only tell us something is wrong somewhere. We can save any hypothesis from doubt by changing our background assumptions. Theories face the test of evidence as wholes. (Kincaid 1996:20)

If we go back to the globe metaphor in Figure 1-2 we can understand more clearly just what is involved here: the contrast among the lines of latitude between the dashes for social science and the solid lines for physical and biological science. Those dashes represent the isolation of hypotheses, concepts, and fields of knowledge from one another, whereas the solid lines indicate relationships among hypotheses, concepts, and fields of knowledge. Quine and Ullian argued that any test of a hypothesis invokes an entire "web of belief." They argued further that this web extends up and down language's levels of abstraction, illustrated by the lines of longitude in Figure 1-2:

> The analytic-synthetic distinction [attempted by logical empiricists] tries to separate the linguistic and factual components behind our beliefs. Some statements are directly tied to confirming evidence or experience; they are synthetic. Other statements gain their credibility from linguistic conventions and thus the empirical data can never refute them; they are accordingly analytic and *a priori*. Quine, however, denied that we could sharply divide evidence in this way, because testing is a holistic affair. (ibid.:19)

Quine and Ullian along with Kincaid thus look to the entire web of solid lines of latitude and longitude depicted in Figure 1-2, implying the great deficiencies in the approach of the social sciences with its longitudes and latitudes made up of dashes.

Kincaid links this web approach to the work of Kuhn, who followed Duhem and Quine and Ullian, and succeeded in expanding their depth and scope. For Kuhn, a great deal is involved in tying hypotheses to the world: metaphysical worldviews, research strategies, standards of good science, interpretations of scientific virtues, and a great deal more. Here, Kuhn succeeds in alerting us, with his concept of paradigm, to a range of structures surrounding a hypothesis. And Kincaid also follows Kuhn in the latter's reaction to postmodernist arguments about the limited worth of the scientific method. Let us recall Kuhn's statement:

> Talk of evidence, of the rationality of claims drawn from it, and of the truth or probability of those claims has been seen as simply the rhetoric behind which the victorious party cloaks its power. (1992:9)

Kincaid's position here is much the same as that of Kuhn: the scientific method can be exceedingly valuable; it is not just another rhetorical device, even if it does not guarantee truth. In Kincaid's words, "empirical or observational evidence can still be the heart of good sense" (ibid:43).

One implication of this web approach to the scientific method has to do with prediction as distinct from explanation. We are presented within much of physical science with formulas that make relatively exact predictions on the basis of a very few variables, such as $F = ma$. It is this very approach that many sociologists have attempted to take over and apply to far more complex situations, using statistics insofar as possible in efforts to apply mathematics to isolated propositions about human behavior. The failure of almost all such efforts to yield much in the way of close relationships by using such isolated propositions should have sent us a warning signal that something was not quite right with this approach. Yet we had no alternative overall approach to the scientific method that would have told us that such efforts at exact prediction should generally be abandoned in favor of an approach that looks to an entire web of factors as well as increasingly better explanations based on those factors. Further, we needed an approach that would have alerted us not to locating factors that could be measured easily and thus thrust into statistical models but rather factors that are theoretically important regardless of how difficult it would be to measure them. With this web approach to social research, however, we no longer have the excuse that present analytic procedures constitute the only game in town.

Another implication of the web approach has to do with its meaning for specialization and subspecialization. Such webs tend to extend far beyond the boundaries of a given subspecialty or specialty, following the ideal of openness within the scientific method. Given the ability of sociology as a whole to extend over fully forty different sections of the American Sociological Association, there is generally little concern with specialization, for most of us tend to be satisfied with the breadth of sociology as a whole. And this view is reinforced by textbooks in the field, which present enormous diversity. Here again we look to the breadth of the discipline as a whole and not to the narrowness of each special field of sociology. If pressed, most of us will grant how little the chapters of our introductory textbooks link to one another and how little communication there actually is among subspecialties, let alone specialties, but our immediate concerns generally are taken up with the progress of our own special field. As a result, our high degree of specialization and subspecialization generally is not seen as a serious problem. We complain about it sometimes, but the lack of any serious alternative to what we are presently doing provides no continuing direction for those complaints. The web approach, however, does offer a genuine direction that contrasts sharply with present procedures.

Yet this approach demands a great deal from us sociologists. Since graduate-school days we have all learned that advancement almost invariably requires a narrow focus so that we can make our mark in some highly specialized field. The further we climb up the ladder the more investment we make in that field and the harder it is to recapture our earlier learning about other fields. If we are indeed to follow this web approach, where are we to gain the confidence in our capacity to move into other fields? And where are we to find the time to do the necessary reading and thinking about those other fields? And where are the exemplars of others who have succeeded in doing this kind of thing? Kincaid of course cannot answer those questions, but he suggests that we no longer use philosophy as a source of final answers:

> We thus have to rethink the relation between philosophy of science and scientific practice. Philosophers have a special place outside of or prior to science itself. Foundationalism—as the idea that philosophers can describe on *a priori* grounds the standards for real scientific knowledge—has to go. . . . The idea that the philosophy of science must be continuous with science itself is widely accepted. (1996:20–21)

It is up to us sociologists to show the way and demonstrate with our own research whether or not a web approach in fact will prove fruitful.

Perhaps what is most significant about Kincaid's work is its timing at the end of the century after all of the failures of the social sciences to live up to the Enlightenment dream. Further, Kincaid has to deal with postmodernist arguments that have challenged not only Enlightenment aspirations but also the credibility of the scientific method. Nevertheless, Kincaid somehow has managed to retain that dream for the social sciences. And beyond that general orientation, he is able to chart a methodological and theoretical direction for the social sciences, one that follows the achievements of the physical and biological sciences, and one that is continuous with Kuhn's basic analysis. But to do so requires that we pay attention to the work of Dewey and Kuhn along with Kincaid. They alert us to the existence of fundamental contradictions between how we see ourselves and how we in fact are, as illustrated in Figure 1-2 by the dashed lines for the social sciences and the solid lines for the physical and biological sciences. It is our own sense of problem as to the adequacy of our discipline that must be at the core of any major change, just as Kuhn has argued about the process by which paradigms are changed. Philosophy can work to give us that sense of problem, not only about our subcultural paradigm within sociology but also about our cultural paradigm or worldview. Yet philosophy can go only so far in charting a direction for the social sciences: it is up to us to carry it much further.

SOCIOLOGY'S PARADIGM

In Chapter 1 we sketched the contrast between a bureaucratic and an interactive paradigm for sociology, centering on our use of the scientific method. Yet our involvement in the former paradigm is so complete, despite our scientific ideals, that a great deal is required if we are ever to shift to the latter. Nietzsche puts forward, metaphorically, the idea of how difficult it is to make fundamental changes in society, just as Kuhn—centering on scientific paradigms—made an argument for such difficulties much later. Nietzsche states:

New struggles.—After Buddha was dead, his shadow was still shown for centuries in a cave—a tremendous, gruesome shadow. God is dead; but given the way of men, there may still be caves for thousands of years in which his shadow will be shown.—And we—we still have to vanquish his shadow, too. ([1887] 1974:167)

Nietzsche's argument in 1887 was that God is already dead, yet any complete change from the worldview where he still lives might take a great many years. Our concern here, however, is not with such complete change, but rather with movement toward such change. Such movement can take place by shifting sociology's paradigm to an extent, and the resulting interaction between that paradigm and our cultural paradigm can—like the back-and-forth motion of a pendulum—yield ever greater changes in both paradigms over time.

Prior to moving to the analysis of our sociological paradigm, it is useful to examine the work of a sociologist who, early in the twentieth century, attempted to gain insights into our cultural paradigm, which he referred to as our *Weltanschauung*. That is a concept that classical sociologists such as Simmel often employed because it opened up the importance of history or historical era for understanding society. A key problem with this concept has been how to conceive of it in a systematic or scientific way. For example, Karl Mannheim writes:

> Is it possible to determine the global outlook of an epoch in an objective, scientific fashion? Or are all characterizations of such a global outlook necessarily empty, gratuitous speculations? . . . Once it is shown that in every cultural product a documentary meaning reflecting a global outlook is given, we have the basic guarantee that *Weltanschauung* and documentary meaning are capable of scientific investigation. . . . The more one is impressed by the inadequacy of explaining *Weltanschauung* in terms of philosophy, the more promising will be the attempt to start from art and analyze all other fields of culture in terms of concepts derived from a study of plastic arts. The "hierarchical level" of plastic art is closer to the sphere of the irrational in which we are interested here. ([1921–22] 1952:9, 45, 51)

Mannheim, like the classical sociologists who preceded him, was struggling to find a direction for sociology that would be both historical and scientific. He saw *Weltanschauung* as a scientific concept partly because he believed that it is possible to take any cultural product whatsoever and then move very far up language's ladder of abstraction so as to find its "documentary meaning" within a global context. Philosophy is inadequate for explaining a *Weltanschauung* because it generally lacks sensitivity to the historical context within which a concrete object occurs. For Mannheim it is essential not merely to have concrete cultural illustrations of a *Weltanschauung* or worldview but also to have a visual basis for it, and this is why he suggests that we look to art for insights. And when he claims that art is close to the "irrational" elements of human behavior, we might recall that perception is tied to human biology and is distinct from our logical or rational modes of thinking and communicating. A *Weltanschauung*, then, is not just based on thought, understanding or communication but is also linked to human biology through perception, just as a self-image is broad enough to extend beyond concepts or ideas to perception. It is in Figure 1-3 that we may trace some links between *Weltanschauung*—or worldview—and other sociological concepts, such as culture and social organization.

David A. Snow

In his 1998 presidential address to the Pacific Sociological Association, David Snow presented a case for the value-added character of sociology: what the discipline adds to the academic experience, things that would be largely lacking if sociology were not alive and well. By centering on our achievements Snow provides an important corrective to one-sided critical assessments both inside and outside our discipline. And his positive view of sociology's achievements can provide a springboard for further achievements relative to the discipline's problems. More specifically, Snow reviews a good deal of sociological literature in order to discover exactly what sociologists have achieved over what has been attained by other disciplines as a basis for confronting downsizing efforts within higher education as well as for clarifying our scholarly and instructional attainments. Snow's systematic and reasoned analysis meshes with the presentation of sociology's conceptual strengths sketched in Figure 1-3. From a critical perspective, however, we can view Snow's arguments as failing to address the balance sheet for sociology relative to larger issues, such as its own Enlightenment aspirations as well as escalating social problems. Snow's purpose did not require pursuit of such topics, yet our own purpose requires that we address them. They point us toward the question of how to build on our present achievements, as presented by Snow, so as to fulfill the Enlightenment promise of sociology.

Snow finds five features that pervade our discipline and that, in their emphasis and importance for scholarship and education, make sociology unique among all other disciplines. He begins with a focus on relational connections, looking to four literatures within the discipline as examples. First, there is the orientation to society as a whole or large social systems, as in the work of Marx, Weber, and Durkheim. Marx, for example, saw institutions like the polity, the family, and religion as linked closely to and revolving around the economic institution, with an emphasis on processes of social stratification like proletarianization. Weber, by contrast, saw the religious institution and culture in general as playing a more independent role in relation to the economic sphere. Granting these differences, both centered on the relationships among forces in society as a whole. Moving up to the present, the world systems approach emphasizes links among the economies of states within a global network, and once again we find stratification as a central force, as in the relationships between developed and underdeveloped economies. Those links involve the phenomenon of dependency, which is central to a second literature that Snow cites: exchange theory. This theory emphasizes not individuals but relationships. Its core thesis is that "an actor is dependent on another to the extent that outcomes valued by the actor are contingent on exchange with the other" (Molm and

Cook 1995:216, quoted in Snow 1999:8; see also Homans 1958; Blau 1964; and Emerson 1962, 1976).

A third literature, following Snow's analysis, is social network analysis, which centers on both the forms and consequences of relationships. Snow goes back to Simmel's *Web of Group Affiliations* (1955) for early work in this area, seeing work in the past two decades as giving us the strategy and tools we require for systematic investigation. He sees examples in a wide variety of areas, as illustrated by the study of collective behavior and social movements (Snow and Oliver 1995:573–75) and the study of corporations and economic forces (Granovetter 1985; Powell and Smith-Doerr 1994). And a fourth literature on relational links has to do with the analysis of face-to-face interaction. Goffman, as a leading early figure who helped to legitimate this area of investigation, maintained that its "proper study . . . is not the individual and his psychology, but rather the syntactical relations among the acts of different persons mutually present to one another" (Goffman 1967:2; quoted in Snow 1999:9). In all four of these literatures Snow finds a focus on relational connections, whether at the micro, the meso, or the macro level of analysis. In his words, "No other social science places such a premium on ferreting out and elaborating the relational bedrock of all of social life" (1999:9).

Before moving to Snow's second distinctive feature of sociology we can begin our own critical analysis. Let us imagine a situation within physics where some physicists did specialized studies of force, others of mass, and still others of acceleration. Someone attempting to review the accomplishments of physics could claim the breadth of the field in that it encompasses force as well as mass and acceleration, analogous to Snow's claim as to sociology's breadth because it includes various specialized studies. Yet how would physicists, given such a situation, ever be able to pull together these specialized studies and come up with the formula $F = ma$? Analogously, if sociologists studying large-scale social structures favor *either* the analysis of social organization *or* the analysis of culture, how can we possibly learn to put the two together—as in Figure 1-1—and take up both social organization and culture simultaneously? Of course, there are great differences between the two fields in the degree of complexity of the phenomena studied, yet those differences make specialization within sociology an even more disastrous affair than it would be within physics. For we have more factors to take into account, as illustrated by the twenty-six concepts depicted in Figure 1-3. Having forty sections of the ASA with little communication among them is practically a guarantee that we will never be able to come up with findings that account for more than a fraction of what we wish to explain.

Snow is apparently unaware that his own analysis implies a direction for the scientific method similar to that outlined in Chapter 1. His master concept, one that is quite abstract, is "sociology." Throughout his article he proceeds to link that abstract concept to other abstract sociological concepts in a rather systematic way, much like the depiction of concepts in Figure 1-3. Thus far in our review, for example, his literatures encompass aspects of social organization, culture, and situational forces. Yet when he cites Goffman's importance for the study of face-to-face interaction, he shows his lack of awareness of the importance of abstract concepts in developing cumulative sociology. Granting Goffman's major substantive contributions, he generally proceeded to invent his own terminology with little regard to linking it to key concepts within the sociological literature. We can applaud him for his insights but certainly not for cumulative development in the area of social interaction specifically or sociology in general. By contrast, Snow develops a masterful analysis of the literatures of sociology, one that relies on its key concepts along with the overall concept of sociology. However, Snow's achievement is by no means the achievements of the literatures he analyzes: they remain isolated from one another and are pulled together primarily in Snow's imagination. Yet that imagination points toward a different way of doing sociology, one where specialists learn to share that imagination.

We might now proceed to the remainder of Snow's analysis of sociology's value-added contributions, reserving further criticism for the conclusion of his analysis, although we cannot include here all of the specific literatures to which Snow refers. Snow's second distinctive focus of sociological analysis is what he calls an emphasis on "contextual embeddedness," referring to the broad context within which any instance of human behavior is located. Here, Snow has an eye for the complexity of human behavior. A molecule is a molecule and can be understood with little reference to its history and links to other molecules, but a human being is different. Snow refers in this connection to Mills's argument that the essence of the sociological imagination resides in the ability to grasp the relationship between individual biography and history in a given society (Snow 1999:9). One literature bearing on the idea of contextual embeddedness centers on the situated character of social action, motive, and meaning. Another focuses on relativity or the idea of comparison.

Snow cites Herbert Blumer's approach to symbolic interactionism as an illustration of a situational orientation, one that by no means neglects social organization and culture but sees them embedded within the situation (Blumer 1969). Snow also cites Scheff's (1997) call for "a parts / whole morphology" as a further illustration of the importance of contextual embeddedness for sociological analysis. That methodological approach—as we shall see in the next section devoted to Scheff—has important implications for changes in sociology's paradigm, as sketched in Chapter 1. As for the embeddedness of motives within the situation, Snow refers to Mills (1940) and Scott and Lyman (1968). The title of Mills's early article, "Situated Actions and Vocabularies of Motive," tells much of the story: motives are situated and cannot be understood apart from their context, and the vocabulary used is a key to the social and cultural forces involved. As for the idea of comparison, Snow sees this applying to the concepts of alienation and "felt deprivation" or, in our own terms, relative deprivation. We might recall the gap between the curves of Figure 1-1, specifying the level of fulfillment *relative to* the level of aspiration. The concepts of alienation and relative deprivation are both linked to that gap.

Snow's third distinctive aspect is sociology's focus on social problems. As examples he cites "gender, race, and ethnic discrimination, substance abuse, drug trafficking, family violence and its various forms of abuse, inequality, poverty, homelessness, juvenile violence and gangs, and environmental degradation" (1999:12). As background to this continuing interest he cites sociology's association with American pragmatism and the Progressive movement as well as the work of the early Chicago sociologists. Of course, we can also go much further back: to the ideals of Auguste Comte, to the commitments of Karl Marx, and to the efforts of Emile Durkheim to alter France's educational institution. As we look to our forty sections within the ASA we might see them as largely split between those emphasizing substantive knowledge (such as organizations, social psychology, comparative/historical, sociology of culture, science, and rational choice) and those emphasizing problems (undergraduate education, medical sociology, crime, peace, environment, sociological practice). Snow insists that although many sociologists believe that a substantive and a problems orientation work against one another, they are in fact mutually supportive. He cites Wilson's *The Truly Disadvantaged* (1987) and Edin and Lein's *Making Ends Meet* (1997), but we can also go back to the work of Karl Marx and, more recently, to the work of C. Wright Mills. Here again, Snow points up a direction for a new sociological paradigm: building bridges connecting these two orientations.

Snow's fourth distinctive aspect of sociology is the discipline's ironic perspective, which involves the denial of well-accepted assumptions. For Snow, "Irony surfaces when things are not as expected or as they should be, when there is an unanticipated mismatch between appearances and reality" (1999:15). He cites Louis Schneider's *The Sociological Way of Looking at the World* (1975), which devotes a section to the topic of irony and its relation to sociology. Schneider writes that irony suggests a discrepancy "between the way things are and they are supposed to be, between promise and fulfillment, between appearance and reality" (1975:xi). We can see this ironic approach illustrated in Figure 1-1, where we tend to accept and remain aware of the cultural values emphasized in the top curve, but we have difficulty becoming aware of the large and apparently growing gap between those values and our ability to fulfill them. Or we can even look to one concept alone, social stratification, for a glimpse into a reality that stands in the way of the fulfillment of many of our promises or values. Figure 1-4 also depicts an ironic situation, where the successes of the physical and biological sciences can turn against us and threaten our very survival. We might look to the concepts within Figure 1-3 as illustrating the relatively invisible tools that sociologists use to uncover contradictions between assumptions and actuality, such as bureaucracy, alienation, and anomie.

Snow has little to say about his fifth distinctive feature of the discipline: its "empirically based and theoretically informed window on the social world" (1999:16). He views this feature as the resultant of the preceding four characteristics of sociology, yielding a change in the student's perception of the human world around him or her. Snow built his essay in an effort to go beyond Randall Collins's "The Sociological Eye and Its Blinders" (1998), for he saw Collins as not explaining exactly what distinguishes the sociological gaze from other perspectives. In fact, this fifth feature of sociology is not merely derivative of the other four, since a way of seeing the world from one moment to the next goes beyond any intellectual understandings to the realm of perception, which has to with our biological structure. We might note that Figure 1-3 includes the concept of biological structure, granting that the focus of sociology has been on other matters. Here, Snow is building bridges not only within areas of sociology but also across disciplines. We might also recall the importance of biological structure in our analysis of worldviews and philosophy at the beginning of the preceding section. There, we suggested the interactive nature of life itself and of biological evolution in particular, indicating that our worldview would do well to take that nature into account. For example, we can learn to see our own interaction with the world around us, versus emphasizing what is external as simply a given with no relation to ourselves.

If Figure 1-3 and Chapter 1 in general suggested the importance of interaction among many components of human behavior—versus our tower-of-Babel situation—then Snow describes literatures that point in the same direction. And if our analysis at the beginning of the preceding section on worldviews suggests the existence of interaction within the physical universe and the importance of evolution and biological structures, then Snow bolsters this view of the centrality of interaction within our universe. All five of Snow's unique features of sociology can be seen as pointing in this same direction. It is not only relational connections among elements from culture and social organization, along with those unearthed by exchange theory and network analysis, which are involved in this approach. It is also the many phenomena that the idea of contextual embeddedness suggests are linked to one another, and this includes the findings of ethnomethodologists no less than symbolic interactionists. This orientation encompasses as well Snow's description of sociology's ironic perspective, which involves the interaction between our taken-for-granted assumptions and our empirical findings. It also includes his view of the importance of interaction between a substantive orientation and a problems orientation within the discipline. And, finally, it includes his link between all of these instances of interaction and our perception of our social world.

Yet without insight into the fundamental gap between our Enlightenment aspirations and our present performance, Snow's analysis becomes little more than a self-congratulatory view of the discipline. There are very good reasons why our credibility both inside and outside the discipline is very low and shows little indication of any improvement. And there are also very good reasons why we have failed and are failing to develop the kind of sociology that is more than minimally cumulative. Granting, as Snow explains, there is good evidence of an interactional orientation *within* many or perhaps almost all of our specialties, that orientation is not carried further so as to link specialties with one another. This is like medical specialists who understand relationships among phenomena *within* a given part of the body but remain unable to link that knowledge with knowledge of other parts of the body. Of course, there is a great deal more to our problems of developing a credible and cumulative sociology than this. For example, there is the interaction between the assumptions or worldview of the researcher—and not just those whom he or she studies—and the researcher's findings. Also, there is the researcher's lack of emphasis on a general set of abstract concepts used throughout the discipline, which carries much of the weight of the knowledge within the discipline. And there is the investigator's overriding focus on accurate prediction to the exclusion of an effort to explain with the aid of a web of concepts that are linked indirectly or directly.

What Snow achieves is a genuine *tour de force* in documenting the unique contributions of sociology. Without any mention of Kuhn he manages to sketch sociology's paradigm, giving us a different take on what was presented in Figure 1-3. Instead of a focus on key sociological concepts he gives us key sociological orientations that point toward culture, social organization, the situation, and the individual. Further, his overall orientation to the centrality of interaction among phenomena points in the direction of the interactive approach to the scientific method outlined in Chapter 1. Of course, to move toward that alternative paradigm we must come to see the deficiencies of our present bureaucratic paradigm with its emphasis on specialization and with little or no interaction outside a given specialty or subspecialty. Yet once we come to understand the limitations of that paradigm we are in a position to move in Snow's direction. By emphasizing the ideals that we all share he sets the stage for such movement.

What we then require in addition is an understanding of our failures to live up to those ideals. For example, we generally fail to carry this interactive approach outside our own specialized fields. Following Kuhn, this would yield awareness of contradictions within our bureaucratic paradigm for the scientific method, yielding a basis for moving toward an alternative paradigm.

A major lack in Snow's approach is the sense of problem and urgency illustrated by John Dewey and sociologists like C. Wright Mills. Snow does indeed defend the importance of an orientation to confronting social problems within the discipline, and he does sustain the claim that this can contribute to our substantive understanding of human behavior. Yet his analysis lacks the implications we can infer from Figures 1-1 and 1-4: that we are rapidly going over the cliff with our accelerating gap between what we want and are able to achieve, and that immediate efforts are essential for any possibility of reversing this trend. To simultaneously see both problems and a direction for solutions is characteristic of an interactive sociological and cultural paradigm. It is one thing to write about such paradigms yet quite another to live them. This is not meant to single out Snow as acting hypocritically. If he is a hypocrite, then so are we all, for we all apparently share a bureaucratic cultural paradigm. Let us recall some of Dewey's words:

> The science that has so far found its way deeply and widely into the actual affairs of human life is partial and incomplete science: competent in respect to physical, and now increasingly to physiological, conditions (as is seen in the recent developments in medicine and public sanitation), but **nonexistent** with respect to matters of supreme significance to man—those which are distinctively of, for, and by, man.

Without recognizing sociology's failures there is no way we can confront them despite all of sociology's past achievements cited by Snow.

Thomas J. Scheff

Over this past decade Thomas Scheff, well-known for his work in the development of a labeling theory of mental illness (see, for example, 1966, 1974), has broken new methodological and theoretical ground. It is one thing to write about a new approach to applying the scientific method to sociology, but it is quite another to apply that method to actual research. Scheff has achieved both with his "part/whole" approach:

> In my earlier volume (Scheff 1990) I specify a general approach to theory and method that I call "part/whole." This approach places equal emphasis on the smallest parts of a social system, the words and gestures in discourse, and

the largest wholes, the institutions that exist within and between nations. In this view, understanding human behavior depends on rapid movement between the parts and wholes, interpreting each in terms of the other.

My approach is similar to what is called morphology in botany, the study of the structure and function of plants. This approach looks at a single specimen in order to understand the species as well as studying the species in order to understand the specimen. Otherwise those details that are needed in explanation might be left out. Darwin's theory of evolution grew out of his observations of extremely small variations in the appearance of species living in separate regions. Had his method been more focused and "rigorous" (by current standards) in the form of an experiment or survey, he probably would have ignored these tiny details.

Applying this method to the human sphere, I have focused on single concrete episodes of behavior: the inception of a marital quarrel at the interpersonal level, two world wars at the international level. I emphasize a "bottom-up" strategy, starting with a detailed examination of single events, as well as a "top-down" strategy, a bird's-eye view of many events in terms of abstract concepts. Part/whole reasoning requires that both strategies be used in conjunction. (1994:4–5)

In a more recent book Scheff (1997) devotes a chapter to explaining the nature of part/whole analysis, and he then proceeds to apply that approach, centering on "the social bond" as illustrated by family relationships. To view Scheff's work systematically, let us link it to the above discussions of Dewey, Kuhn, Kincaid, and Snow, looking to his accomplishments along with what remains to be done. Dewey, writing after World War I as well as after World War II, was deeply concerned with the failures of the social sciences to give us the understanding of social problems that we desperately and urgently needed. Half a century later those needs appear to be even more pressing. We can tell even from the title of Scheff's *Bloody Revenge: Emotions, Nationalism, and War* that he is most interested in applying his approach to important social problems. That book addresses both family problems and problems of war, analyzing in the latter case the origins of World War I and World War II. And Scheff does not shy away from making specific recommendations as to how we might confront those problems more effectively. Here, Scheff illustrates Snow's analysis of sociology as a discipline that emphasizes concern for confronting social problems. Further, Scheff's theoretical and methodological contributions make it quite clear that this concern not only does not inhibit substantive contributions but can actually strengthen those contributions.

Like Dewey it is not just social problems that are Scheff's concern: it is also our approach to social research that he takes to task. For example, he sees human behavior as more complex than social scientists generally assume: "It is clear that societies (and the human relationships which constitute them) ride upon extraordinarily complex processes" (1997:1). By homing in on research methods in the social sciences, Scheff once again follows Dewey's deep concern with the failures of the social sciences to yield the understandings we require for solving basic problems. Scheff's part/ whole analysis is an effort to give us an alternative approach to scientific research. Scheff's commitment here has extended over an entire decade and has involved a number of books and studies. Our illustrations in Chapters 3 and 4 of sociological research that follows the interactive paradigm sketched in Chapter 1 will dip heavily into Scheff's work. Here, we might simply suggest that his work shows a deep commitment both to substantive knowledge and to the development of social technologies built on that knowledge. Actually, things work in the other direction as well: his efforts to build such technologies have yielded important substantive knowledge. A key focus of his is on the importance yet relative neglect by the social sciences of emotions. This goes back to a research paradigm that generally requires clear and exact measurements. Yet emotions are intangible and difficult to measure.

This recent work by Scheff speaks directly to Kuhn's belief in the importance of scientific paradigms and, by extension, our own concern with cultural paradigms and worldviews. In his concluding chapter Scheff uses the term "habitus" to invoke what Kuhn suggests by "paradigm," drawing heavily on the work of the anthropologist Clifford Geertz (see, for example, Geertz 1973, 1983). We have, then, both a disciplinary habitus and a cultural habitus, with this word emphasizing what we take for granted yet that remains largely invisible or unexamined. Scheff comments on our cultural habitus:

> The first four qualities of commonsense reality [cultural habitus] that Geertz names—naturalness, practicalness, thinness, and accessibleness . . . might be subsumed by a more abstract concept: cultural systems of commonsense are *non-reflexive*. . . . The system of commonsense operates outside of awareness. An outsider can reflect and comment on it, but an insider cannot. . . . Freud . . . had the temerity to argue, in a society that was still sufficiently traditional that he might have been lynched, that religion functioned as a mechanism of defense. The belief in an after-life, particularly, is a defense against the fear of dying. In modern societies, religion has lost much of its sacredness, but commonsense has not. (1997:224)

Scheff goes on to examine the disciplinary habitus of the social sciences, corresponding to Kuhn's concept of scientific paradigm. For example, he sees the discipline of economics as requiring that research be quantitative and—insofar as possible—expressed by mathematical models. He introduces Richard Feynman's essay "Cargo Cult Science" (1986) as a possible explanation for such behavior. When the U.S. military abandoned the islands of the South Seas after World War II, many islanders felt deprived of the "largesse" they had received during the war. In an effort to bring back our armed forces they formed cults that set up runways with improvised lights with the hope that the planes would return. Feynman criticized academic psychologists, claiming that their imitations of the physical sciences in an effort to bring scientific status to their discipline missed the point of how those sciences actually work, much like the islanders missed the point of what had brought the military to the South Seas. Those psychologists were imitating procedures but failing to understand the basic rationale or logic involved in scientific research, just as we might view the habitus of economists as equally missing the point of scientific research. Scheff suggests that all of the social sciences generally function as cargo cults in their adulation of quantitative procedures. Of course, it is far easier for us sociologists to see psychologists and economists as creating cargo cults than to see ourselves in this way.

What we have in Scheff's work is the combination of a highly critical with a highly constructive approach, corresponding to what Kuhn suggested as the basis for scientific revolutions as well as what was sketched in Chapter 1 as the basis for an interactive versus a bureaucratic scientific method. His cargo-cult metaphor carries along with it a very serious criticism of sociology and social science, yet he also proceeds to give us a direction for moving beyond such superficiality. For one thing, his focus on context carries forward what Herbert Blumer was attempting to achieve with symbolic interactionism as well as what is of concern to ethnomethodologists in general. Whereas work in these areas often ignores or only gives lip service to macro social structures, Scheff welcomes such structures into his analysis. In this way he comes closer to Kincaid's view of a web approach to the scientific method. Scheff's use of a hermeneutic orientation where sociologists do secondary analyses of something previously published in order to extend understanding of the context involved provides a direction for following that web approach by integrating diverse specialized orientations within the discipline. This is a theme that is important for Snow, and it is one that Kincaid's work points toward. Yet what Scheff does is far more than talk about its importance: he demonstrates in his own research how we might move in this direction.

Scheff ends his book on a realistic note that gives recognition to the enormous task facing sociologists. He quotes from one analyst of academia: "Each tribe has a name and a territory, settles its own affairs, goes to war with the others, has a distinct language or at least a distinct dialect and a variety of ways of demonstrating its apartness from others" (Bailey 1977). And he also quotes from Geertz, whose analysis can be applied both to our scientific and our cultural paradigms:

The problem of integration of cultural life becomes one of making it possible for people inhabiting different worlds to have a genuine, and reciprocal impact upon one another. . . . The first step is surely to accept the depth of the differences; the second to understand what these differences are; and the third is to construct some sort of *vocabulary* in which they can be publicly formulated. (Geertz 1983; emphasis added by Scheff)

Although Scheff suggests in a tentative way that part/whole analysis might provide the common vocabulary that is needed, this quote implies our distance from achieving that common vocabulary. Given our Tower of Babel with forty different sections in sociology, how far have we come toward taking Geertz's steps of accepting, understanding, and communicating the nature of our differences, let alone building bridges based on what we have in common? And if this is our situation within sociology, how much more difficult is it outside academia?

Yet following Kuhn it is exactly this kind of orientation to the depth of our problems that is fundamental to a scientific method that requires awareness of problems as a basis for addressing them. We might at this point look not to the achievements of Scheff along with Snow, Kincaid, Kuhn, and Dewey but rather to where they fall short. For one thing, they are limited in giving us a vocabulary of concepts that could become the basis for a sociology that is rapidly cumulative, granting that Snow and Scheff move us a considerable distance in this direction. Snow is able to point up theoretical orientations that, in the opinion of most sociologists, have yielded a great deal of insight into human behavior. He has, then, opened up areas of convergence. Scheff's critiques of our standard methodology where we do very little in learning about the context of a given finding opens up what could become areas of convergence but in fact remain problems and areas of divergence. Yet both Snow and Scheff fail to focus on a set of abstract concepts that most of us might agree point up areas of convergence. Snow's orientations fail to take much note of the great divergence of our forty sections and the fact that we specialists speak very different dialects. Scheff gives us a highly credible demonstration of an approach to research yielding both substantive and applied insights, yet it lacks a set of concepts that could become the basis for ever more convergence among our forty sections.

Scheff's work shares with Dewey's work a sense of problem and urgency at this time in history. Dewey, writing just after World War I as well as World War II, believed deeply in the importance of reconstructing our cultural and scientific paradigms. And Scheff's efforts to understand the forces that led to World War I and World War II parallels Dewey's efforts. Yet here we are hardly past the close of the century and the millennium, where the gap between aspirations and fulfillment throughout the world shows every sign of not only being much wider than in the eras of those wars but also appears to be accelerating, and modern societies apparently lack the sense of urgency let alone the commitment that Dewey and Scheff illustrate. If we look to Figure 1-1, what we see is a rapidly accelerating gap between aspirations and fulfillment. And if we proceed to analyze the forces behind that gap, as in Figure 1-4, what we see is the deep involvement of the fundamental structures of modern society. These include our worldview along with our basic patterns of culture and social organization, and that gap is also linked to relatively invisible problems like anomie, alienation, and addiction. What appears to be required in this situation is a widespread sense of problem and urgency that equals that of Dewey and Scheff, for we are apparently faced with an escalating yet invisible crisis in modern society.

Up to this point we have sketched in Chapters 1 and 2 a contrast between a bureaucratic and interactive scientific method along with a bureaucratic and interactive cultural paradigm. And following Kuhn we have also sketched an approach—following an interactive scientific method—for changing from one paradigm to another. Given our focus on putting forward a way of doing research that differs markedly from present techniques, and given our additional focus on a cultural paradigm that differs markedly from our present one, we have not been able to present a case for the utility of this new approach. In one sense, as we have learned from recent critiques of the scientific method, we can rely on no scientific approach with any certainty that it will yield truth or even that it will produce successive approximations to truth. In another sense, a major part of the determination of utility has to do with whether or not a method will yield both substantive insights, and also whether effective problem-solving technologies can be built on the basis of the understandings that emerge from the approach, and this remains largely a task for the future.

Let us bear in mind that our focus in this book is on reconstructing the scientific method we sociologists use, granting that reconstruction necessarily also involves the cultural paradigm and worldview within which our scientific method is embedded. Our illustration applying an interactive scientific method to the problem of the invisible crisis of modern society is just that: an illustration. It is in Part II that we shall attempt to strengthen that illustration, yet that does not change our overall focus on the scientific method. The illustration suggests the utility of our approach to the scientific method, yet it may ultimately prove to be wanting. What is crucial for that approach is not what appears in this book but what occurs afterward, when that method is used again and again.

PART II

ILLUSTRATING THE WEB APPROACH TO THE SCIENTIFIC METHOD

Our focus here is on the same fundamental problem of modern society that was portrayed in Figures 1-1 and 1-4: the accelerating gap between aspirations and fulfillment, or between cultural values and patterns of social organization affecting their fulfillment. We can see this general and highly abstract problem linked to a variety of relatively invisible as well as visible problems. As for the former, Durkheim's concept of anomie points toward the failure of society to provide the kinds of norms that succeed in guiding the individual toward fulfilling values or goals. Another example of a relatively invisible problem in modern society is suggested by Marx's concept of alienation, referring to the individual's persisting feelings of isolation from others, from the physical environment, from his or her own productive activities and even from his or her own "species being" or biological structure. Also, we have the phenomenon of social stratification, which is viewed as far less tolerable within modern society—with its emphasis on the cultural value of equality—than it was in the past. Yet another illustration of an invisible problem is the phenomenon of relative deprivation—roughly similar to feelings of jealousy—associated with the impact of stratification and cultural values like that of equality within a situational context.

These relatively invisible problems are not the ones we see in the media, yet they are linked to our more familiar social problems. For example, Durkheim linked anomie to suicide and Merton related it to a variety of problems, including crime, addiction, and other forms of deviant behavior. We can relate the phenomenon of alienation to various kinds of mental problems and, more generally, to such problems as divorce within the family, absenteeism at work, and lack of voting or participation in community affairs. As for social stratification, we have the gap between the rich and the poor both internationally and within wealthy nations, the undemocratic concentration of power in the hands of a few throughout modern societies, and persisting patterns of racism, sexism, ageism, classism, and ethnocentrism. And we see relative deprivation illustrated by situational acts of discrimination and labeling or prejudice, whether toward others or self. Of course, there are also a great many other relatively visible problems tied to those relatively invisible ones, such as war, terrorism, illiteracy, physiological problems, and spousal and child abuse.

It is in Chapter 3 of Part II that we strengthen our understanding of the fundamental problems confronting modern society, as illustrated within Figures 1-1 and 1-4. How deeply are we enmeshed in those problems and how threatening are they? How compelling is the evidence for their continuing escalation? And how weighty is that evidence in comparison to assessments based on traditional and specialized research procedure? At the same time we gain further understanding of procedures for pulling together this evidence based on the web approach to the scientific method presented in Part I. We use this approach to gain insight into our basic problems, centering on anomie, alienation, social stratification, and relative deprivation. In Chapter 4 our focus is on paths toward solutions rather than on uncovering problems, following the approach sketched in Figure 1-5. For example, can we actually learn alternatives to change our patterns of racism, classism, sexism, ageism, ethnocentrism? As we shall see, continuing awareness of the depth of our problems is crucial to the development of directions toward solutions. Yet that awareness is affected by the relative invisibility of those problems as well as forces like our present worldview, which encourage us to bury our heads in the sand.

These chapters do no more than begin a process of integrating sociological knowledge based on that web approach to the scientific method. To the degree that this process continues it should yield a solid platform of knowledge on which we will be able to build highly effective social technologies. That knowledge will not be so accurate as to yield precise predictions from causes to effects, as we have experienced within the physical sciences. Nevertheless, it should be able to carry more and more of the weight of our knowledge of human behavior, carrying our understanding far beyond the knowledge on which specialized experts depend. In Chapters 5 and 6 we shall examine some of the implications of this situation for sociology as well as for modern society, freeing ourselves to speculate as to the impact of the fulfillment of Comte's Enlightenment dream. For example, how might each and every individual learn to apply the range of available knowledge from the social sciences to his or her problems of everyday life? In that kind of world, what kind of an economy would we develop? What would happen to social stratification and bureaucracy? to our worldview and cultural paradigm? What would the world be like if the problem of the survival of the human race were behind us and the development of our human potential became our fundamental problem?

A fundamental element of our approach in Chapters 3 and 4 will be the abstract sociological concepts presented in Figure 1-3 and discussed in

Chapters 1 and 2. To emphasize their importance they will appear in bold-face throughout these two chapters and the ones to follow. This will not be done every time they appear, as that would be too distracting for the reader, but only whenever boldface can make some contribution to our understanding. To the reader who is quite familiar with those concepts, even this might appear to be a species of overkill, converting those concepts into clichés and detracting attention from concrete description. Yet if indeed we are all largely prisoners of a bureaucratic worldview, then we will have learned to use all language—and not just the language of sociology—in a simplistic way. More precisely, our vernacular usage will not take into account the range of forces involved within any given situation, as illustrated by the range of concepts within Figure 1-3. Presenting those concepts in boldface can serve as a reminder of that complexity not only for sociological concepts but for all of our language. It is, then, a reminder that appears to be useful not only for learning a broader approach to the scientific method but also for learning to step out of the bureaucratic worldview that stands in the way of learning that broader approach.

It is one thing for us sociologists to talk about scientific ideals like this web approach to the scientific method, yet quite another thing for us to actually employ that orientation in our work. For what we do in our work depends on our way of thinking in everyday life, which in turn is linked to our worldview and cultural paradigm. Presenting key sociological concepts in boldface will by no means alter our way of thinking in sociology, much less our way of thinking in everyday life, our worldview, and our cultural paradigm. Nevertheless, it appears to be a useful step in that direction, provided that we do not quickly pass over those concepts and ignore the significance of their appearance. Our approach here is similar to that taken by Alfred Korzybski (1933), an engineer who invented "general semantics" as a movement designed to apply the scientific method to problems of language and human communication (see also Hayakawa 1949; and the journal ETC). Korzybski tried to teach people "consciousness of abstracting," or momentary awareness that one's concepts or verbal maps are not identical with the territory they supposedly portray, and he used physical devices to serve as reminders of this difference between map and territory. Our use of boldface might similarly serve as a reminder that our concepts are far from the broad territory suggested by the full range of concepts in Figure 1-3, and that reminder should help us to move closer to that territory.

Such use of boldface for basic sociological concepts might serve to remind us as well of the importance of abstract concepts in sociological research. In our present usage of the scientific method within sociology we tend to see such abstract concepts as no more than a prelude to empirical research, where it is only concrete data at a low level of abstraction that is truly the stuff of science. This is quite understandable, given sociologists' interpretation of the scientific method as centering on what empirical research has that philosophical speculation does not have. Yet it is our abstract sociological concepts that serve to organize and carry the weight of our many pieces of concrete evidence. By presenting them in boldface we give credit where credit is due, for they constitute our basic tools for understanding human behavior. And insofar as they invoke our many concrete studies, when we are able to link those abstract concepts to one another we simultaneously integrate the vast number of concrete studies on which they are based. And as a result, we increase the credibility of analyses that employ those abstract concepts manyfold. Granting that those analyses do not yield exact predictions, nevertheless they go far beyond our achievements with highly specialized analyses.

One further point about using boldface for those concepts is the optimism this conveys about the possibilities for fundamental change within society and the individual. If those concepts are indeed our fundamental tools and if there are a very limited number of them, then the task of learning to understand them should not be very difficult. And this would give us a beginning on a path toward using them in everyday life and, as a result, challenging the everyday thought that is deeply embedded within our worldview and cultural paradigm. What is involved here is not the acquisition of a Ph.D. or the reading of hundreds of books or waiting decades for the younger generation to start from a different perspective. But we should remain aware that learning to understand those concepts—and this includes their interrelationships—is no more than a first step on the road to learning to use them in everyday life. It is this latter task that is truly difficult. For what it requires among other things is that our everyday usage of concepts in thought, feelings, and action be accompanied by the same kind of consciousness that this use of boldface points to. Granting that there is no biological barrier to our developing such consciousness from one moment to the next, there are many other barriers that must be overcome.

Mills's idea of the sociological imagination as an ability that all of us can learn to develop points us in this direction, granting that he did not give us a procedure for how to develop that imagination. Let us recall his description of that way of thinking:

> That imagination is the capacity to shift from one perspective to another—from the political to the psychological; from examination of a single family to comparative assessments of the national budgets of the world; from the theological school to the military establishment; from considerations of an oil industry to studies of contemporary poetry. (1959:7)

Can we afford to be optimistic about the possibility that sociologists, let alone everyone else, can develop a sociological imagination? Perhaps the more relevant question is: Can we afford to be pessimistic about this possibility? If we give credence to Figures 1-1, 1-2, and 1-4 it is the latter question that is most relevant. For we all appear to be caught in a world situation that requires nothing less than the replacement of our narrow worldview with this broader one. And learning to apply this system of concepts to the literature of sociology and to everyday life appears to be a procedure for that replacement.

3

The Web Approach Illustrated:
The Invisible Crisis of Modern Society

Just what is the weight of evidence for the existence of an invisible crisis in modern society, as depicted in Figures 1-1 and 1-4? Given the breadth of those figures, our traditional specialized approach to the scientific method cannot help us in answering this question. For example, the two curves in Figure 1-1 depict **cultural values** and patterns of **social organization** that limit their fulfillment, yet our traditional procedures generally separate studies of culture and social organization. However, we shall in fact be able to take both curves into account simultaneously with the web approach to the scientific method described in Chapters 1 and 2. That approach is based on the range of sociological concepts depicted in Figure 1-3, which include culture, social organization, the situation, and the individual. However, since fully twenty-six concepts are involved, for the sake of clarity we shall focus on four concepts that represent those four areas, with the freedom to link those concepts with others. We shall select the four that point up relatively invisible problems, as suggested in the introduction to Part II: **anomie, alienation**, social **stratification,** and **relative deprivation**. These relatively invisible problems are tied to a wide range of relatively visible ones, as outlined above.

Our aim in this chapter is to use our web approach as a basis for pulling together a good deal of sociological research that bears on the existence of an invisible crisis in modern society, using studies that bear on anomie, alienation, social stratification, and relative deprivation as specific ways to assess the existence of that crisis. For example, to the extent that we find **anomie** and **alienation** to be increasing as modernization proceeds, that will constitute partial evidence for the existence of that crisis. In this way we shall also be illustrating the utility of that web approach to the scientific method for integrating available sociological knowledge and bringing it to bear on any given problem, by contrast with the utility of the present-day specialized approach, which achieves little integration. At the same time we shall be illustrating the utility of this web approach for gaining insight into a given problem, insight that generally would not emerge from traditional specialized procedures. For example, can we learn from this approach more about how anomie and alienation derive in part from the individual's behavior in specific situations? Conversely, can we also learn how the existence of anomie in society and alienation in the individual operate as forces tending to produce certain kinds of behavior within a given scene?

ANOMIE

Most of us are familiar with Durkheim's description of increasing rates of suicide during recessions, when the "means" for fulfilling "needs" no longer work very well and a wide gap results between what many need and are in fact able to obtain. And we are equally familiar with his conclusion that suicide rates increase in times of prosperity, when many raise their aspirations beyond what they might reasonably expect to achieve. Yet we are hardly familiar with the implications for our discipline of Durkheim's research for a very wide range of other findings throughout the discipline. The concept of **anomie** can usefully encompass both aspects of **culture** and **social organization** simultaneously, forces that are represented by the two curves in Figure 1-1. Anomie is defined in the glossary as the failure of society's norms or rules to guide the individual's actions toward the fulfillment of values or interests. Such fulfillment or lack of it is in turn largely based on patterns of **social stratification** and **bureaucracy**. This approach points to the very heart of the structural emphasis of our discipline, where structure is seen as going beyond social organization so as to include culture. Even our contemporary belated emphasis on culture generally fails to point up the importance of this kind of breadth.

Beyond this breadth is Durkheim's dynamic analysis of the relationship between **anomie** and the continuing process of industrialization, an orientation to social and cultural change that points directly at Figure 1-1:

> If anomie never appeared except, as in the above instances, in intermittent spurts and acute crisis, it might cause the social suicide-rate to vary from time to time, but it would not be a regular, constant factor. In one sphere of social life, however—the sphere of trade and industry—it is actually in a chronic state.
>
> For a whole century, economic progress has mainly consisted in freeing industrial relations from all regulation. . . . First, the influence of religion was felt alike by workers and masters, the poor and the rich. . . . Temporal power, in turn, restrained the scope of economic functions by its supremacy over them and by the relatively subordinate role assigned them. Finally, within the business world proper, the occupational groups by regulating salaries, the price of products and production itself, indirectly fixed the average level

of income on which needs are partially based by the very force of circumstance. . . . Actually, religion has lost most of its power. And government, instead of regulating economic life, has become its tool and servant. . . . Even the purely utilitarian regulation of them [the appetites] exercised by the industrial world itself through the medium of occupational groups has been unable to persist. Ultimately, this liberation of desires has been made worse by the very development of industry and the almost infinite extension of the market . . .

Such is the source of the excitement predominating in this part of society, and which has thence extended to the other parts. There, the state of crisis and anomy is constant and, so to speak, normal. From top to bottom of the ladder, greed is aroused without knowing where to find ultimate foothold. Nothing can calm it, since its goal is far beyond all it can attain . . . The wise man, knowing how to enjoy achieved results without having constantly to replace them with others, finds in them an attachment to life in the hour of difficulty. But the man who has always pinned all his hopes on the future and lived with his eyes fixed upon it, has nothing in the past as a comfort against the present's afflictions. ([1897] 1966)

Durkheim goes on to cite data for eight different countries or areas: France, Switzerland, Italy, Prussia, Bavaria, Belgium, Wurttemberg, and Saxony, comparing suicide rates for individuals in trade and industry with those for agriculture. In every case with the exception of Wurttemberg— where the suicide rate for those in trade still exceeded substantially that for agriculture—the rates for trade were more than double that for agriculture. In Italy the rate for trade was ten times that for agriculture, and for Bavaria it was triple that for agriculture. As for rates in industrial versus agricultural occupations, the differences were similar although not as marked, although in the case of Belgium the rates were the same and in the case of Wurttemberg there was a slight reversal of the general trend. Durkheim concludes that "Anomy, therefore, is a regular and specific factor in suicide in our modern societies" (ibid.:258).

Granting that Durkheim is able to present only some data for a limited number of countries, and granting that we have learned since his time about many of the flaws in that data—such as his lack of an independent measurement of anomie—nevertheless his analysis points clearly in the direction of a trend toward **anomie** in modern societies, a trend linked to the very process of industrialization or modernization. And given his theoretical analysis, which we have updated by linking it to patterns of culture and social organization, such increasing anomie stems from the fundamental **social structure** of modern society. Durkheim himself cites changes in three major **institutions** encouraging such trends: religion, the political institution, and the economic institution. If we turn to Figure 1-1 we can see the close relationship between Durkheim's argument and those two curves as well as the rapidly increasing gap between them, which suggests increasing anomie. The top curve has to do with the "liberation of desires [which] has been made worse by the very development of industry and the almost infinite extension of the market." The bottom curve has to do with the individual's realistic opportunities for fulfilling those desires or appetites. Durkheim's analysis implies the question of how far these trends can go without the end of society as we know it.

Durkheim's *Suicide* is a hard act for any sociologist to follow, yet Max Weber's *The Protestant Ethic and the Spirit of Capitalism* ([1905] 1958) manages to fill the bill. For example, he helps us to understand just where the top curve in Figure 1-1—the revolution of rising expectations—comes from. Our focus here is on the bridge Weber managed to build between two major **institutions**: the economic institution and religion, a bridge that yields insight into our revolution of rising expectations and its associated **cultural values.** At the risk of greatly simplifying Weber's argument, he saw the Protestant ethic as channeling the individual's enormous motivation for attaining religious salvation into a work ethic, which became the basis for "the spirit of capitalism," as illustrated by this quote from Benjamin Franklin:

> Remember, that *time* is money. He that can earn ten shillings a day by his labour, and goes abroad, or sits idle, one half of that day, though he spends but sixpence during his diversion or idleness, ought not to reckon *that* the only expense; he has really spent, or rather thrown away, five shillings besides. . . . Remember, that money is of the prolific, generating nature. Money can beget money, and its offspring can beget more, and so on. . . . He that kills a breeding-sow, destroys all her offspring to the thousandth generation. He that murders a crown, destroys all that it might have produced, even scores of pounds. . . . The most trifling actions that affect a man's credit are to be regarded. The sound of your hammer at five in the morning, or eight at night, heard by a creditor, makes him easy six months longer; but if he sees you at a billiard-table, or hears your voice at a tavern, when you should be at work, he sends for his money the next day; demands it, before he can receive it, in a lump. (cited in Weber [1905] 1958:48–49)

Robin Williams's (1970; see also 1992) analysis of major value orientations in American society—most of which can be generalized to modern society as a whole—helps us to locate the particular **cultural values** associated with this work ethic. If we are indeed to pay serious attention to the phenomenon of **culture** as a powerful structure and not simply employ this concept vaguely and merely in passing, then it is essential that we identify its basic elements. This procedure is a kind of "socioanalysis" paralleling a psychoanalysis of the individual: we locate some of the invisible and internal forces that we denizens of modern society all generally share, bringing them up to the light of day. For example, Williams cites "achievement and success," as illustrated by our reverence for our thrifty, hardworking, ambitious, and successful rail splitter who became our most revered president. There is also the cultural value of "activity and work," illustrated by our Puritan tradition and Protestant ethic. And there is a focus on the related cultural values of "efficiency and practicality," "material comfort," and economic and technological "progress." In amassing evidence for his analysis, Williams went beyond citing a few illustrations to a review of the literature on our cultural history.

Max Weber's quote from Benjamin Franklin illustrates an understanding that reaches beyond the forces within culture so as to include the momentary situation, anticipating current work in such theoretical fields as ethnomethodology, symbolic interactionism, and rational choice theory. The examples Franklin cites, such as "he that goes abroad," "the sound of your hammer at five in the morning," and "if he sees you at a billiard-table" call to mind the situational concepts listed in Figure 1-3. For example, each of these examples suggests either **conformity** to or **deviance** from norms tied to the cultural values associated with the work ethic. Further, the hammer and billiard-table examples illustrate the phenomenon of **social interaction.** And that in turn influences a creditor to **label** an individual as either a good credit risk or a bad one. In the former case we may speak of positive **reinforcement** for the borrower, and in the latter case negative **reinforcement.** All of this depends on just how the creditor interprets what he hears and sees, that is, his **definition of the situation**, something that is largely invisible or at least most difficult to measure. If in fact he "sends for his money the next day" the borrower might well come to feel **relative deprivation**, that is, feelings of unjustified loss relative to other borrowers who have not suffered the same fate.

Our analysis of the cultural values and situational behavior associated with Protestantism should be extended backward to include the early Judeo-Christian tradition if we are understand more of the range of **cultural values** in modern society that are involved within our "revolution of rising expectations." The upper curve for the revolution of rising expectations depicted in Figure 1-1 begins with preindustrial society and not industrial society, where it gathers momentum. We must, then, go back to preindustrial society to examine its roots. Here we might turn initially to a passage from Genesis describing God's work during the sixth and last day of the creation:

God said, "Let us make man in our own image, in the likeness of ourselves, and let them be masters of the fish of the sea, the birds of heaven, the cattle, all the wild beasts and all the reptiles that crawl upon the earth." God created man in the image of himself, in the image of God he created him, male and female he created them. (Gen.1:26, 27; *Jerusalem Bible* 1966:6)

What this passage reveals is what Williams has called the **cultural value** of "individual personality," referring to the ultimate worth of every individual. What could be more worthy than a creature fashioned in the very image of God? It is a worth that is beyond the worth of all other living creatures, for the human being is to be the master of them all. And this creation also suggests the **cultural value** of "equality," for all human beings—regardless of their station in life—are equally required to conform to the dictates of the Ten Commandments. Closely related to equality and individual personality is the **cultural value** of "freedom" from the arbitrary use of power by others. We are familiar with this concept in early modern times in relation to the struggles for freedom during the French and American revolution, with the French slogan "liberty, equality, fraternity" and the focus in the Declaration of Independence on life, liberty, and the pursuit of happiness. We can also go back to the biblical Ten Commandments for the beginnings of this same idea of freedom. Not only is the individual protected from the arbitrary taking of his life, family relationships, and property but also from the abuse of his neighbor.

As we saw in Figure 1-1, it is when we integrate knowledge of cultural values with knowledge of opportunities for fulfilling them that we begin to see some of the deepest problems of modern society, and it is just here that Robert Merton's essay "Social Structure and Anomie" (1949) yields insight. His focus there is not on suicide as in Durkheim's analysis, but on crime, along with other kinds of **deviance**:

> It is, indeed, my central hypothesis that aberrant behavior may be regarded sociologically as a symptom of dissociation between culturally prescribed aspirations and socially structured avenues for realizing these aspirations. . . . With such differential emphases upon goals and institutional procedures, the latter may be so vitiated by the stress on goals as to have the behavior of many individuals limited only by considerations of technical expediency. . . . As this process of attenuation continues, the society becomes unstable and there develops what Durkheim called "**anomie.**" . . .
>
> Several researches have shown that specialized areas of vice and crime constitute a "normal response" to a situation where the cultural emphasis upon pecuniary success has been absorbed, but where there is little access to conventional and legitimate means for becoming successful. . . . Recourse to legitimate channels for "getting in the money" is limited by a class structure which is not fully open at each level to men of good capacity. Despite our persisting open-class ideology, advance toward the success-goal is relatively rare and notably difficult for those armed with little formal education and few economic resources. The dominant pressure leads toward the gradual attenuation of legitimate, but by and large ineffectual, strivings and the increasing use of illegitimate, but more or less effective, expedients. . . . It is only when a system of cultural values extols, virtually above all else, certain *common* success-goals *for the population at large* while the social structure rig-

orously restricts or completely closes access to approved modes of reaching these goals *for a considerable part of the same population*, that deviant behavior ensues on a large scale. (pp. 128, 136–37)

What Merton gives us is direct support for a key aspect of Figure 1-1, the importance of taking into account both the top curve and the bottom curve simultaneously, and he links the relationship between those curves to the phenomenon of **anomie**. Centering on work-related **cultural values** like Williams's "material comfort" or "achievement and success," combined with what he implies as patterns of **social stratification** that limit access to the fulfillment of those cultural values, he sees the result as anomie coupled with **deviance** such as crime. Poverty or social stratification alone is not enough. Also, poverty found in the midst of plenty is not enough. Rather, it is poverty or social stratification combined with "common success goals for the population at large"—or cultural values that remain unfulfilled due to social stratification—that is central to anomie and the genesis of deviance like crime. We can extend Merton's analysis to other cultural values than work-related ones, such as the people-oriented values of "individual personality" and "freedom." Since these are also cultural values, this suggests the existence of anomie affecting the rich as well as the poor, for they too are limited in their ability to fulfill those values.

As for Figure 1-1's portrayal of increasing **anomie** along with modernization, Merton adds to Durkheim's and Weber's evidence with his analysis of the rise of science in seventeenth-century England:

In the Puritan ethos . . . in contrast to medieval rationalism, reason is deemed subservient and auxiliary to empiricism. . . . It is on this point probably that Puritanism and the scientific temper are in most salient agreement, for the combination of rationalism and empiricism which is so pronounced in the Puritan ethic forms the essence of the spirit of modern science. . . . And one of the consequences of Puritanism was the reshaping of the social structure in such fashion as to bring esteem to science. ([1938] 1996:231–32)

Let us grant Durkheim's analysis of anomie as associated with industrial occupations, and let us also grant Weber's view of the Protestant ethic as a key basis for the engine driving the industrial revolution, an engine also associated with our revolution of rising expectations. In addition, we have Merton's analysis of that ethic as also central to the engine driving the physical and biological sciences along with their associated technologies. The result, as we see in Figure 1-1, is an ever-widening gap between aspirations and the ability to fulfill those **cultural values** throughout modern society.

Durkheim's and Merton's analyses of anomie continue to be central to current sociological research (Crutchfield 1992), illustrated by studies attempting to explain the high crime rates in the United States by invoking our cultural value of equality coupled with patterns of social stratification. We can also look outside the sociological literature for analyses of **anomie**. For example, the neo-Freudian analyst Karen Horney's *The Neurotic Personality of Our Time* (1937) views widespread neuroses in American society as deriving from basic contradictions within our **social structure**:

> When we remember that in every neurosis there are contradictory tendencies which the neurotic is unable to reconcile, the question arises as to whether there are not likewise certain definite contradictions in our **culture**, which underlie the typical neurotic conflicts . . .
>
> The first contradiction to be mentioned is that between competition and success on the one hand, and brotherly love and humility on the other. On the one hand everything is done to spur us toward success, which means that we must be not only assertive but aggressive, able to push others out of the way. On the other hand we are deeply imbued with Christian ideals which declare that it is selfish to want anything for ourselves, that we should be humble, turn the other cheek, be yielding. For this contradiction there are only two solutions within the normal range: to take one of these strivings seriously and discard the other; or to take both seriously with the result that the **individual** is seriously inhibited in both directions.
>
> The second contradiction is that between the stimulation of our needs and our factual frustrations in satisfying them. For economic reasons needs are constantly being stimulated in our culture by such means as advertisements, "conspicuous consumption," the ideal of "keeping up with the Joneses." For the great majority, however, the actual fulfillment of these needs is closely restricted. The psychic consequence for the **individual** is a constant discrepancy between his desires and their fulfillment.
>
> Another contradiction exists between the alleged freedom of the **individual** and all his factual limitations. The individual is told by society that he is free, independent, can decide his life according to his own free will; "the great game of life" is open to him, and he can get what he wants if he is efficient and energetic. In actual fact, for the majority of people all these possibilities are limited. What has been said facetiously of the contradiction between the alleged freedom of the individual and all his factual limitations can well be extended to life in general—choosing and succeeding in an occupation, choosing ways of recreation, choosing a mate. The result for the individual is a wavering between a feeling of boundless power in determining his own fate and a feeling of entire helplessness. (pp. 287–88)

In our own terms, Horney's first contradiction is one between work-related **cultural values** like "achievement and success" and people-oriented **cultural values** like "equality" and "individual personality." If we refer to Figure 1-1, Horney suggests that there is a contradiction between the work-related and people-oriented cultural values located within the top curve. From our own perspective this is yet another source of **anomie** in modern society as a whole, since that contradiction makes it difficult for all of us to fulfill both work-related and people-oriented values. Her second contradiction has to do with the gap between the two curves of Figure 1-1 and parallels the analyses of Durkheim and Merton. Her third contradiction yields further illustrations of that gap by looking to the broad **cultural values** of "freedom" and "individual personality" and finding sharp limits on our ability to fulfill such values. The title of Horney's book, *The Neurotic Personality of Our Time*, suggests her view of these contradictions as a modern phenomenon, supporting the temporal orientation of Figure 1-1. In her analysis Horney follows Durkheim's and Merton's location of these basic contradictions within social structure, following sociologists' location of anomie within social structure.

We might look again at Horney's first contradiction, that between cultural values, and her statement as to our alternatives:

> For this contradiction there are only two solutions within the normal range: to take one of these strivings seriously and discard the other; or to take both seriously with the result that the individual is seriously inhibited in both directions. (ibid.: 287)

A third alternative, which is outside "the normal range," also can be applied to the second and third contradictions. It is to reject our seesaw or zero-sum **worldview** and point toward an alternative **worldview** as well as an alternative **culture** or cultural paradigm that supports that alternative worldview. Merton saw such a possibility when he distinguished five possible reactions to the gap depicted in Figure 1-1: **conformity, deviance,** ritualism, retreatism, and "rebellion," with the latter constituting that third alternative. For Merton this involves both a rejection of existing goals or cultural values as well as existing means for their fulfillment coupled with the development of alternative goals and means. We shall have a good deal more to say about this third alternative in Chapter 4.

We might also examine more closely what we achieve by drawing on the work of a psychoanalyst as distinct from a sociologist. Horney is able to bring to bear on our sociological concepts the results of a great deal of experience having to do with the nature of mental problems, adding to the credibility of our own analysis. Further, her analysis yields further insight into our own. For example, her third contradiction—between the alleged freedom of the individual and all his factual limitations—suggests that we take into account our own concepts of **biological structure** and **physical structure**. Specifically, she states that we are unable to choose our own parents, and we can add that our genetic makeup, race, and ultimate death also are not under our control, suggesting the impact of biological structure. Further, we cannot control our time and place of origin, such as our era and initial nationality, suggesting the impact of the physical structures of time and space. Yet to the extent that we emphasize the cultural value of "freedom" we tend to ignore such biological and physical limitations. Also, our bureaucratic worldview and culture or cultural paradigm lead us to focus on a very narrow range of phenomena, again distracting our attention from our limitations and thus encouraging further anomie.

ALIENATION

By keeping the concept of **anomie** distinct from that of **alienation** we are able to take into account the power of two structures and not just one: **social structure** *and* the **individual**. This is exactly what Horney achieved when she distinguished between culture and the neuroses of the individual. Some empirical studies based on survey analysis have attempted to measure **anomie** solely through measuring individual responses in interviews or questionnaires. For example, an early study by Mizruchi (1960), based on 618 interviews derived from a probability sample of adults in a small upstate New York city of 20,000, suggests the tenor of such research procedures. Mizruchi used Srole's scale of "anomia" (1956) along with Chapin's Social Participation Scale (1952), finding support for Bell's (1957), Meier and Bell's (1959), and Srole's (1956) findings of an inverse relationship between class and "anomia." In our terms, "anomia" might best be classified as a species of **alienation** rather than **anomie**, since its focus is on the individual and not on social structure. From this perspective all of these studies found a relationship between social stratification and alienation in a society where the cultural value of equality is exceedingly important.

Let us look to Marx's own discussion of **alienation,** the basis for our definition of the concept as "persisting feelings of isolation from self, others, one's own biological structure and the physical universe." It was during his last years on the Continent, prior to the Revolution of 1848 in Paris, that he penned these words:

> We have now considered the act of **alienation** of practical human activity, labour, from two aspects: (1) the relationship of the worker to the *product of labour* as an alien object which dominates him ... (2) the relationship of labour to the *act of production within labour.* This is the relationship of the worker to his own activity as something alien and not belonging to him. ... This is *self-alienation* as against the above-mentioned **alienation** of the *thing.* ... Since alienated labour: (1) alienates nature from man; and (2) alienates man from himself, from his own active function, his life activity; so it alienates him from (3) the species. ... For labour, *life activity, productive* life, now appear to man only as *means* for the satisfaction of a need, the need to main-

tain his physical existence . . . free, conscious activity is the species-character of human beings. (4) A direct consequence of the **alienation** of man from the product of his labour, from his life activity and from his species-life, is that *man* is *alienated* from other *men*. ([1844] 1964, 125–27, 129)

Alienation centers on the feelings of the **individual** just as does **relative deprivation**, but it appears most useful to treat it as a structure within the **individual** rather than as a situational occurrence, following Marx's analysis of its far-reaching impact. Granting its location within the individual, it is *also* a product of forces within society and history. With this focus on the individual we do not take away from the importance of those larger forces. Further, with this definition of alienation in terms of "persisting" feelings, we come to see this phenomenon as a structure no less than culture and social organization. Marx saw the worker's **alienation** as tied to the nature of industrial society, where the proletariat is a pawn of the capitalist system. In particular, Marx saw the **social stratification** between capitalists and workers as basic to generating the worker's alienation. As a result of his experiences within the workplace as a pawn of the bourgeoisie, the worker is dehumanized (**biological structure**), has no control of his own activities (**personality structure**), comes to be divorced from his physical environment (**physical structure**), and also loses out on relating to his fellow man (**social structure**). Marx's breadth here suggested our own definition of the **individual** as "a system of social, personality, biological, and physical structures."

Marx is explicit in his essay on **alienation** about the worker's worsening situation as industrialization proceeds:

The object produced by labor, its product, now stands opposed to it as an *alien being*, as a *power independent* of the producer. The product of labor is labor which has been embodied in an object and turned into a physical thing; this product is an *objectification* of labor. The performance of work is at the same time its objectification. The performance of work appears in the sphere of political economy as a *vitiation* of the worker, objectification as a *loss* and as *servitude to the object*, and appropriation as *alienation* . . . .It is just the same as in religion. The more of himself man attributes to God the less he has left in himself . . .

The more the worker produces the less he has to consume; the more value he creates the more worthless he becomes; the more refined his product the more crude and misshapen the worker; the more civilized the product the more barbarous the worker; the more powerful the work the more feeble the worker; the more the work manifests intelligence the more the worker declines in intelligence and becomes a slave of nature. . . . Labor certainly produces marvels for the rich but it produces privation for the worker. It produces palaces, but hovels for the worker. It produces beauty, but deformity for the worker. It replaces labor by machinery, but it casts some of the workers back into a barbarous kind of work and turns the others into machines. It produces intelligence, but also stupidity and cretinism for the workers. ([1844] 1964:122–24)

Just as Figure 1-1 depicts an increasing gap between aspirations and fulfillment as we move from preindustrial to modern society, so does Marx claim that **alienation** continues to increase as industrialization proceeds. Fritz Pappenheim, a student of economics, sociology, and philosophy, uses different words than ours about that increasing gap to argue for increasing alienation from preindustrial to modern society:

> In the present stage of history man has means of self-realization at his command which were unknown to him in former periods. The immense advance in science and technology has helped him to understand the forces of nature to such a degree that he is not any longer at their mercy. . . . Once this concept of the individual's sovereignty has been awakened in the minds of men, a new climate is prepared. The consciousness that man's yearning for self-realization is thwarted becomes a crushing experience which could not have existed in previous stages. (1959:114–15)

If that increasing gap is a basis for anomie within social structure—as depicted in Figure 1-1—then there is good reason to believe that it is also a basis for alienation within the individual.

Marx's interest in the importance of the concept of **alienation** did not subside after his early analyses of the idea in his *Economic and Philosophical Manuscripts of 1844*, despite many arguments that have been made to the contrary. There is ample evidence from his writings that he continued to be concerned about the problem of alienation throughout the remainder of his life and that he continued to use this concept (Meszaros 1970:217–53). What is at stake here is whether or not the body of Marx's mature work does in fact support his early interest in alienation. And more generally, what is also at stake is whether we need to concern ourselves less with structures within the **individual**, limiting ourselves largely to concerns with **social structure**. Even more generally, if Marx did in fact alter his emphasis, should that be decisive in influencing us to turn away from the importance of the individual in addition to social structure, following the emphasis of sociology? The approach taken here is that a focus on either **social structure** or the **individual**—regardless of what Marx did or did not do—makes little sense if we wish to pull together the findings of sociology and also open up to the full range of what we've learned from the social sciences.

Much of current research has moved from Marx's focus on the **alienation** of the worker to a concern with the **alienation** of the voter, with several analysts maintaining an interest in the forces within social structure that produce political alienation. For example, Seeman (1975) has argued that feelings of powerless involved within alienation are linked to political contexts within Western democracies, where it is difficult for the individual to influence governmental decisions. Lipset and Raab (1978) tie alienation with its sense of powerlessness and meaninglessness to anomie, where the political rules of the game are no longer effective in accomplishing what they are supposed to accomplish. In a broad view of the implications of nonvoting within American politics, Bowles and Gintis (1987) look to fundamental political and economic contradictions within American life and point up the efforts of businesses to stifle reform. Such continuing concerns with alienation by contemporary researchers (see, for example, Lo 1992) extending beyond the economic to the political institution support an argument for the increasing prevalence of this phenomenon in contemporary society.

Georg Simmel gives indirect support to Marx's emphasis on the importance of alienation in modern society with his focus on the problems of the **individual**:

> The deepest problems of modern life flow from the attempt of the individual to maintain the independence and individuality of his existence against the sovereign powers of society, against the weight of the historical heritage and the external culture and technique of life. ([1903] 1971:324)

Simmel's orientation to the individual is no less broad than that of Marx. And for Simmel—who maintained his focus on the individual throughout his life—the problem of maintaining "independence and individuality" is *the* fundamental problem of modern society. For example, Simmel saw the modern metropolis with all of the "fluctuations and discontinuities of the external milieu" as pushing the individual to protect his inner emotional life by surrounding it with a hard shell of intellectuality. Such emotional repression was, for Simmel, not characteristic of earlier times with its "smaller circle in which the inevitable knowledge of individual characteristics produces . . . an emotional tone in conduct, a sphere which is beyond the mere objective weighting of tasks performed and payments made" (ibid.:327).

Later in his essay on "The Metropolis and Mental Life" Simmel takes up the effect of "the money economy" and its impact on the individual, just as he does in great detail in his *The Philosophy of Money*. One point he makes on the importance of watches and punctuality within the money economy reveals the general approach he adopts throughout all of his writings:

> But here too there emerge those conclusions which are in general the whole task of this discussion, namely, that every event, however restricted to this superficial level it may appear, comes immediately into contact w

depths of the soul, and that the most banal externalities are, in the last analysis, bound up with the final decisions concerning the meaning and the style of life. (ibid.:328)

What we have here is Simmel's claim that situational behavior, like the **definition of the situation** in a given scene as requiring punctuality, is closely linked to the basic **social structure** of the money economy. Indeed, his approach meshes with our own web orientation to the scientific method, where all of the social and individual structures are tied to all of the situational concepts, which in turn are tied to our apparently trivial momentary behavior.

Still later in that essay Simmel gives us a preview of a key idea he was to emphasize a few years later about a growing gap between "objective culture," following the usual definition of culture, and "subjective **culture**," or the development of the individual's subjective life:

This discrepancy is in essence the result of the success of the growing division of labor. For it is this which requires from the individual an ever more one-sided type of achievement which, at its highest point, often permits his personality as a whole to fall into neglect. In any case this overgrowth of objective culture has been less and less satisfactory for the individual. . . . He is reduced to a negligible quantity. He becomes a single cog as over against the vast overwhelming organization of things and forces which gradually take out of his hands everything connected with progress, spirituality and value. (ibid.:337)

If we look to Figure 1-1 once again, then we might see the increasing gap between the two curves as depicting not only patterns of objective culture (top curve) and its fulfillment as affected by social organization (bottom curve) but also objective culture (top curve) and subjective culture (bottom curve). The **individual** is being ripped apart by modernization no less than is **social structure**.

Simmel also redefines the top curve of that figure so as to focus in particular on the revolution of rising expectations in its relation to the development of the **individual**, contrasting the eighteenth with the nineteenth centuries:

The eighteenth century found the individual in the grip of powerful bonds which had become meaningless—bonds of a political, agrarian, guild and religious nature—delimitations which imposed . . . an unjust inequality. In this situation arose the cry for freedom and equality. . . . Alongside of this liberalistic ideal there grew up in the nineteenth century from Goethe and the Romantics, on the one hand, and from the economic division of labor, on the other, the further tendency, namely, that individuals who had been liberated from their historical bonds sought now to distinguish themselves from one another. No longer was it the "general human quality" in every individual but rather his qualitative uniqueness and irreplaceability that now became the criteria of his value. (ibid.:338–39)

If we now see the top curve of Figure 1-1 as centering only on the **cultural value** of "individual personality" and not on the full range of cultural values, then we can see this figure as supporting Simmel's view on "the deepest problems of modern life."

If the four major classical figures contributing to the origins of sociology were Marx, Durkheim, Simmel, and Weber, it would be useful to examine once again the work of Weber, this time within the context of the phenomenon of alienation. For it was Weber more than any other sociologist, classical or modern, who has given us insight into the nature of **bureaucracy,** and he has also uncovered fundamental problems associated with the change from preindustrial social organization to our modern bureaucratic society. A key distinction Weber makes in his examination of modern social organization is that between "formal rationality" and "substantive rationality":

A system of economic activity will be called "formally" rational according to the degree in which provision for needs . . . is capable of being expressed in numerical, calculable terms, and is so expressed. . . . Expression [of the concept] in money terms yields the highest degree of formal calculability. . . . On the other hand, the concept of substantive rationality . . . [implies] that it is not sufficient to consider only the purely formal fact that calculations are being made. . . . In addition, it is necessary to take account of the fact that economic activity is oriented to ultimate ends of some kind, whether they be ethical, political, utilitarian, hedonistic, the attainment of social distinction, of social equality, or of anything else. (1964:185)

This distinction poses a problem for the individual, paralleling Simmel's argument about the viability of "subjective culture" in the face of our rapidly expanding "objective culture." In a world where our money economy and bureaucratic modes of organization have been expanding rapidly, we have an increasing emphasis on formal rationality. In this kind of world, what happens to substantive rationality? In other words, what happens to the range of our **cultural values,** which cannot easily be reduced to quantitative calculation, such as "individual personality," "equality," "freedom," and "democracy"? Weber maintained that formal rationality was a cornerstone of the developing industrial society, but does this also imply that within such a society those cultural values are threatened, as Simmel argued in relation to "subjective culture" versus "objective culture"? Weber noted a continuing trend toward the "rationalization" of society, seeing this as delivering to all of us a far more efficient economy and a far more rational—in the formal sense—society than what had been experienced within preindustrial society. Here we can all cite the vast improvements associated with modern science and technology and our modern legal system. But must we pay as a price the loss of substantive rationality?

Weber links his argument about the Protestant ethic to this rationalization of society and raises a similar question at the end of his *The Protestant Ethic and the Spirit of Capitalism*:

> The Puritan wanted to work in a calling; we are forced to do so. For when asceticism was carried out of monastic cells into everyday life, and began to dominate worldly morality, it did its part in building the tremendous cosmos of the modern economic order. This order is now bound to the technical and economic conditions of machine production which today determine the lives of all the individuals who are born into this mechanism, not only those directly concerned with economic acquisition, with irresistible force. Perhaps it will so determine them until the last ton of fossilized coal is burnt. In Baxter's view the care for external goods should only lie on the shoulders of the "saint like a light cloak, which can be thrown aside at any moment." But fate decreed that the cloak should become an iron cage . . .
>
> No one knows who will live in this cage in the future, or whether at the end of this tremendous development entirely new prophets will arise, or there will be a great rebirth of old ideas and ideals, or, if neither, mechanized petrification, embellished with a sort of convulsive self-importance. For of the last stage of this cultural development, it might well be truly said: "Specialists without spirit, sensualists without heart; this nullity imagines that it has attained a level of civilization never before achieved." (pp. 181–82)

We can see here that Weber's argument that modernization poses enormous problems for the individual parallels that of Marx and Simmel and points to increasing **alienation**. For if alienation has to do with the feelings of the individual, then we are surely moving toward alienation when we become "specialists without spirit, sensualists without heart" and when we find ourselves located within an "iron cage" from which we cannot escape. Weber himself was not able to point to a direction for resolving these problems, just as we might claim that neither Marx nor Simmel offered a viable solution. Following the schematic diagram within Figure 1-1, the problem of alienation is increasing exponentially along with that of anomie. But is there evidence from modern sociology that this is in fact occurring? If we look to the process of automation and the development of a wired world as examples of what is happening within the modern economy, some sociologists expressed hope several decades ago that automation would reduce alienation by relieving the worker of repetitive and mindless activities, freeing him or her for more creative work (Bell 1973; Blauner 1964; Shepard 1971). But others have argued more recently that the reverse in fact appears to be the case (Erikson 1986).

Erikson reviews a number of analyses in his examination of what automation has actually yielded (Braverman 1974; Burawoy 1979; Edwards 1979; Glenn and Feldberg 1977; Feldberg and Glenn 1983; Noble 1984; Wallace and Kalleberg 1982). What automation appears to require is not the kind of "mastery of materials" and "maturity of judgment" that was characteristic of the crafts in preindustrial times but rather "a deftness of hand, a sureness of eye, a quickness of reflex." This is far from the kind of creativity we like to think of as characteristic of craftsmanship. For example, the computer can store the information about how to deal with problems and procedures that has been laboriously developed by many thousands of workers over many years of work. This information can then be fed to any worker as needed, and that worker can simply learn to depend on that information rather than exercise any kind of personal creativity. Further, the computer can be programmed so as to control to a significant degree the moment-to-moment activities of the worker, yielding a result that is analogous to the control of the worker's actions on the assembly line. Instead of allowing for more creativity, it would appear that the computer generally would reduce creativity substantially.

Yet there is a good deal more to this story of the impact of the computer and automation on the worker. For one thing, there is "the boredom that comes from doing almost nothing at all," replacing the boredom associated with doing repetitive tasks with the boredom associated with tending the machine. Machines have moved beyond those completely in the control of the worker, such as the hammer or the bulldozer. They can be programmed to have their own self-correcting devices and their own sets of requirements, becoming the worker's master rather than a servant. Another problem is that automation takes the individual far away from closeness to the materials needed for work. For example, instead of feeling the texture of leather or the grain of wood, the worker in continuous process plants never sees or touches the raw materials or finished products. And he or she often is divorced from the rhythms of family or community life, serving instead the rhythm of the evening shift required by the plant. If we focus on **social stratification,** then much of what has occurred with automation is greatly increased control by management over the life of the worker, as elaborated by a number of studies (Kohn 1976; Kohn and Schooler 1983; Mortimer and Lorence 1979; Mottaz 1981; Walsh 1982).

Erikson concludes his analysis with questions about the impact of alienation on the rest of the individual's overall existence, assuming that this phenomenon is indeed a powerful force in the workplace:

The work of Melvin Seeman in particular gives us a secure place to stand when we consider the anatomy of alienation (see 1959, 1972, 1975, and 1983 in particular). But being raised now is the question of what it does to the human spirit in other ways. Do the conditions that Marx encouraged us to think of as alienating add in any appreciable way to the sum of human indifference, brutality, exhaustion, cruelty, numbness? Is there any relationship between alienation and the passion with which capital punishment is promoted, insults to national honor resented, people of other kinds demeaned? I have no idea. I only know that such questions are important, sympathetic, and, in principle answerable. (1986:7)

Erikson suggests here that although we know something about alienation in the workplace—illustrated by the work of Seeman—we sociologists are largely ignorant of its impact within society as a whole. And he suggests that this kind of knowledge is "important, sympathetic, and, in principle answerable."

Yet are we in fact so ignorant? Within what has been called a bureaucratic approach to the scientific method, as supported by a bureaucratic **worldview** and cultural paradigm, we are indeed ignorant of the impact of alienation on society. But within the web approach to the scientific method and the interactive **worldview** behind it, we know a great deal about alienation's impact from the existing sociological literatures. For example, once we define **social stratification** in a sufficiently abstract way, then it extends far beyond hierarchies within the workplace so as to encompass all institutions and the full range of our experiences in society. And if we have discovered that stratification is a powerful force for the production of alienation within the workplace, then we have simultaneously discovered—indirectly—stratification to be a powerful force for alienation outside the workplace. Further, if social stratification and alienation are both linked to a bureaucratic worldview and cultural paradigm, then those phenomena in turn link alienation to "the passion with which capital punishment is promoted, insults to national honor resented, [and] people of other kinds demeaned." The literature on which Figure 1-3's twenty-six concepts rests, and the relationships among those concepts, carry us far beyond the workplace and into society as well as the individual.

SOCIAL STRATIFICATION

If we examine the bottom curve of Figure 1-1, which depicts our opportunities for fulfilling cultural values, perhaps the central finding of sociology as a whole is the existence of **social stratification**—a species of social organization—in all aspects of modern life, whether it be in the form of classism, ethnocentrism, racism, sexism, ageism, or other isms that have received less attention. These patterns appear to be as old as human history, as suggested by Gideon Sjoberg's analysis of *The Preindustrial City* (1965) as well as by the classic debate between Davis and Moore (1945) and Tumin (1953). Although Davis and Moore and Tumin all agreed on the existence of stratification in all known societies, Tumin claimed that stratification is not inevitable within society. The Davis-Moore-Tumin debate reflects sociological interest in stratification following the early work of Marx and later work of Weber. We have for example Weber's (1958) extension of the idea of class stratification to "status" and "party", Dahrendorf's (1959) emphasis on power as a key to stratification, Lenski's (1966) focus on power and privilege throughout history, and Collins's efforts to link microstratification with macrostratification: in "wealth, politics, careers, families, clubs, communities, lifestyles" (1975:49), and a wealth of current analyses of classes and elites along with classical work other than that cited above (Etzioni-Halevy 1997).

Marx, Weber, and Simmel were all deeply concerned with the impact on the individual of patterns of social organization that limit the opportunities of the individual, such as social stratification and bureaucracy. There was Marx's focus on the effect of social stratification on the worker within the workplace, Weber's concern with the individual's loss of "spirit" or "heart" within our bureaucratized society, and Simmel's view of the "sovereign powers of society" and the "technique of life" as crushing the individual. We have of course numerous studies showing increasing opportunities for upward mobility accompanying the change from preindustrial to modern society (Ries 1992), and this is illustrated by the upward movement of the bottom curve. Yet that upward movement is limited, as illustrated by continuing worldwide patterns of stratification in most areas of life. To emphasize that upward movement and to ignore the persistence of social stratification in all areas of life is to fail to make visible invisible forces like the impact of our **worldview** and cultural paradigm, of formal rationality and of objective culture. As a result, it is to remain imprisoned by them, committing ourselves as individuals with our enormous potential to an iron cage from which there is no escape.

It is only when we view the top and bottom curves of Figure 1-1 simultaneously and in relation to long-term historical change and also introduce the concepts of **anomie** and **alienation** that we can see the trend and significance of an increasing gap between those curves. What is involved here is nothing less than long-term historical patterns pitting the two fundamental aspects of social structure, culture, and social organization, against one another, with anomie and alienation summarizing our modern situation. And from Simmel's perspective, we can also interpret the top curve to represent "subjective culture" and the bottom one to represent "objective culture," with the gap posing a threat to individuality. Here Simmel is concerned with how **culture** threatens the **individual**, following the strengthening and development of aspirations for individuality during the nineteenth century. Durkheim and Weber were also much concerned with the power of culture over the individual, foreshadowing the increasing emphasis within sociology on language and culture in the twentieth century. This triumvirate along with many contemporary sociologists are extending the idea of social stratification far outside the workplace, but we must be open to moving beyond a fixation on early Marxist theory to understand what is being accomplished.

Albert Bergesen (1993) has maintained that this recent emphasis on language and culture has penetrated deeply into Marxist theory, serving to broaden that theory just as the work of Weber and many others not directly dealing with Marxist theory have accomplished. For example, he cites the work of Antonio Gramsci as turning Marx on his head by looking to what Marx saw as the superstructure of culture and treating it as more fundamental than what Marx saw as the substructure of the political economy with its patterns of social organization:

> In Gramsci, then, culture / ideology is no longer a thing to be explained but is now a thing that does the explaining; no longer an effect, it is becoming a cause. Social consciousness, as world views, is becoming the central factor in both the perpetuation, and change, of social relations. Rule and revolt now hinge on belief. The class struggle becomes the ideological struggle, Gramsci's "war of position" between bourgeois and proletarian world views, as they struggle for the mind of the working class to convince them they best represent the interests of the social formation as a whole. It is important to note that it is not only that the realm of ideas is the locus of struggle, but that struggle itself is becoming ideological struggle, taking precedence over more overt political and class struggle. Control over the structure of social consciousness is becoming the prerequisite for control over the structure of production. (ibid.:5)

Gramsci does not give up on Marx's emphasis on the importance of social stratification, but he emphasizes the operation of stratification within the realm of culture in general and ideology in particular. Domination or "hegemony" comes to be linked to power within the sphere of culture. For example, power comes to be based on knowledge, with a new class emerging: the professionals and intellectuals with their advanced degrees. Gramsci distinguishes, however, between "traditional intellectuals," such as journalists, men of letters, philosophers, and artists, with the "new intellectuals" closely associated with the rise of science and technology:

> The mode of being of the new intellectual can no longer consist in eloquence, which is an exterior and momentary mover of feelings and passions, but in

active participation in practical life, as constructor, organiser, "permanent persuader" and not just a simple orator. (1971:5–18)

What is crucial in Gramsci's approach is not any proof that the old forms of social stratification have been replaced by newer ones: there is indeed little evidence for that thesis. Rather, it is his focus on the importance of new kinds of social stratification within the sphere of culture.

To the extent that this thesis is correct, then all of us become subordinate to professional "experts" in all spheres of life, and we can even consider them as a not-so-new class, as is argued in this contemporary study:

> Power based on knowledge is, we contend, a basic form of class power. Modern-day experts are only the latest in a long succession of specialists who have spun knowledge into gold—in every age and every part of the world. Knowledge-based hierarchies long antedate those based on ownership of capital and are just as essential to understanding power in human societies.
>
> In the world's greatest ancient civilization, bureaucratic scholar-officials or "mandarins" ruled China for over 1,000 years. They contended that, according to the laws of nature, "there should be two kinds of people: the educated who ruled and the uneducated who were the ruled." The mandarins created a formal class hierarchy based on Confucian credentials conferred by exams. Many other groups have, in different ways, also built great power from knowledge claims, including, as we shall show, witch doctors in tribal societies, priests in the Middle Ages, and organized craftsmen in nineteenth-century capitalist societies.
>
> Today's most powerful knowledge class—professionals—does not rule in any society. But professionals have infused both capitalism and socialism with a modern mandarin logic. By creating a belief in their own knowledge as objective expertise, and helping to organize schooling and the division of labor to suit their own ends, professionals have essentially turned modern knowledge into private property. As in mandarin China, such intellectual property is becoming the coin of the realm, convertible into class power, privilege, and status. (Derber, Schwartz, and Magrass 1990:4–5)

To the extent that the argument that we have stratification throughout the realm of culture is correct, then this directly affects profoundly the bottom curve of Figure 1-1, regardless of whether or not this new class is now dominant in society. The question of who rules society is not crucial for our own analysis, but rather the question of the size of the gap in that figure and whether that gap is growing. If we all are learning to bow down to the supposed wisdom of professional experts in all areas of life, then this has serious repercussions on cultural values such as equality and individual personality. The existence of an expert society is also the existence of a stratified society within the various realms of culture. What we require at this point are examples of how such stratification works and what the repercussions are on the individual, and to this end we turn to radio and tele-vision. Does it appear, for example, that our mass culture and expert soci-ety set up forces that encourage **anomie** and **alienation**? What kinds of **social stratification** are involved? Are there trends that suggest increasing anomie, alienation, and social stratification? Is there evidence of a bureau-cratic approach to the scientific method? of a bureaucratic **worldview** and cultural paradigm?

Central to fulfilling such **cultural values** as "individual personality," "equality," and "freedom," and equally important for our sustaining a **worldview** within which we see ourselves as achieving intimate **social interaction** within a close-knit **group,** is the nature of our sexual relation-ships. This is an area of our lives that has remained largely hidden throughout much of Western history, yet phone-in radio programs like Dr. Ruth Westheimer's "Sexually Speaking" and television programs like "Good Sex with Dr. Ruth," renamed "The Dr. Ruth Show" in 1985, have helped to change this situation, not to speak of widespread changes in ad-vertising and popular culture. Here we can compare ourselves with oth-ers not with respect to income, education, occupation, or status but with respect to our sexual performance. How well are we doing on a hierarchy with "good sex" at the top and "bad sex" at the bottom? Marc LaFountain proceeds to analyze these shows from the perspective of Michel Foucault, whose contributions—which include a three-volume work on the his-tory of sexuality—have been central to the development of postmod-ernist ideas (see, for example, 1972, 1977, 1978, 1984a, 1984b). Foucault has been deeply concerned with, among other things, sexual repression in the modern world along with the forces within culture that result in this re-pression.

LaFountain sees Dr. Ruth "as one who perpetuates the 'repressive hy-pothesis,' which ironically spreads biotechnical power and domination in the name of liberating individuals":

> Dr. Ruth . . . discusses sex as a natural, biological process that, because of "scrupulousness, an overly acute sense of sin, of hypocrisy" (Foucault 1978, pp. 128-129), has been silenced, censored, and repressed. . . . This speech is the crucial, initial move in deproblematizing and normalizing sex. What Dr. Ruth advocates is awareness, discussion, enlightenment, liberation, and, most of all, "good sex" . . .
>
> . . . It can even be advanced that Dr. Ruth's popularity may be understood as a form of aggrandizement of the Master that stems from the therapeutic process of transference. What appears as liberating is arguably little more than the promotion of a celebrity and a media form at the expense of those who become dependent on her for perspective and renewal. Dr. Ruth is not simply Dr. Ruth. The spread of the repressive hypothesis is as much a trans-ference of power as is the growth of dependence and security a function of

psychotherapy. Domination is often propagated in the name of emancipation. (1989:129, 135)

Following LaFountain's argument, we can see Dr. Ruth's programs as fostering the phenomena of anomie, alienation, and social stratification. With respect to **anomie** or the failure of cultural norms to help the individual to fulfill cultural values, the mass appeal of these programs—which open up problems relating to orgasm, masturbation, perverse pleasure, impotence, and frigidity—is an indication that the guidance that society presently gives us for our sexual conduct is inadequate. As for **alienation,** our sexual failures illustrate an extremely deep species of alienation that is biological as well as social, and that is crucial to the individual's self-image. LaFountain also claims that Dr. Ruth illustrates the "repressive hypothesis," engaging in "domination" or **social stratification** in relation to her viewers, teaching them dependence on her expert views on sexuality and depriving them of their own autonomy in this vital area. From a historical perspective, we can see Dr. Ruth's programs as extending the forces of social stratification and alienation to new and vital areas of our lives, very far beyond the kind of alienation which Marx wrote about in relation to the worker in the workplace. And, equally, we can see the extension of the phenomenon of anomie beyond Durkheim's analyses.

If Dr. Ruth gives us an example of social technology on radio and TV, let us now examine a nonfictional as well as a fictional portrayal of science on TV. We shall take up two studies of TV programs on science and technology, one centering on "NOVA", the award-winning PBS documentary science series, and the other on science fiction on prime-time television and in particular on "Star Trek" and "Star Trek: The Next Generation." Our focus on science and technology stems from the centrality of those forces for the development of the modern world with their patterns of **social stratification** and bureaucratic **worldview**. Susanna Hornig (1990) has analyzed television's "NOVA" with a focus on the nature of its presentation of modern science and the scientist. Following the emphasis on "constructionism" within contemporary sociology, her analysis is headed "Television's NOVA and the Construction of Scientific Truth." She views a TV program as not merely a passive conductor of ideas but rather as an active constructor of what we understand to be the nature of science and the scientist. This follows our own emphasis on the importance of the situation in the development of social structure as well as the individual.

Hornig's investigation includes a detailed analysis of "The Race for the Superconductor," an episode describing the efforts of researchers from Sweden, Japan, and the United States to create a compound that will conduct electricity while losing no power to electrical resistance. Hornig prefaces her analysis by pointing up the popular view that "a broad distribution of knowledge is important to a democratic society." She goes on to describe some of the constraints imposed on "NOVA," such as its need to secure funding from government, the corporate world, and the general public. There is also the necessity of obtaining the cooperation of scientists, the support of the scientific community, and the interest of the audience.

"The Race for the Superconductor" begins with a picture of a cube that appears to be dancing in the air, and this picture is repeated at intervals throughout the presentation. A narrator tells us that superconductors will have an "almost magical power," and he later summarizes: "It would be like the world of Buck Rogers come to life with this super-electricity." And later the audience is told that "airplanes could be flown merely by the pilot's own thoughts," although it is not informed just how this could be accomplished.

There are interviews with researchers in Zurich, Tokyo, and the United States interspersed with "shots of complex laboratory equipment, chemicals in jars, blackboards complete with equations, and various explanatory models and diagrams, including another periodic chart from which elements appear to take off and fly toward the viewer" (ibid.:15). We are shown bottles of liquids dripping through complicated arrangements of tubes, although there is little or no explanation of the purposes behind such research. Scientists are portrayed as a breed apart from lab assistants and ordinary people. They generally appear either in a suit and tie or a laboratory coat, talking in front of blackboards covered with equations and in offices filled with books and piles of papers containing their notes. The episode presents "The Race for the Superconductor" as a competitive race among researchers in Sweden, Japan, and the United States, with the suggestion that the stakes are no less than world domination. The enemy appears to be Japan, and if the United States "gets there" first, Japan may still win because of its systematic procedures for developing industrial applications for new knowledge. Researchers are described as workers in the "trenches of superconductors."

The narrator proclaims that a race is on for "technological supremacy." And there is a close tie between science and scientific technology, on the one hand, and economic development, on the other. The show begins with a group of men celebrating the inauguration of a new Silicon Valley company, and the formation of this company frames the entire show. What is at stake is not only technological supremacy but also economic prosperity. What is claimed here is not the prosperity of certain corporations along with their shareholders and managers but the greater good of all. No distinction is made "between the use of science by private interests in pursuit of economic gain and the use of science in the service of society" (ibid.:20). Overall, scientific development is presented as being driven by the desire for economic progress, with no mention of conflicts between private economic interests and broader societal interests. In this episode science and technology are equated with physical science and technology, with biological science in the background, represented by some squirming DNA molecules. Social science and its technologies remain completely invisible.

Hornig compares "The Race for the Superconductor" with other episodes, such as "The Hidden Power of Plants," and she finds substantial differences in the way different scientists are treated:

> Diagrams and models used in the superconductor show are surrounded by a black background and often accompanied by music that reinforces the suggestion of mystery. Diagrams and models in the medicinal plant show, on the other hand, are superimposed over a pale green leaf, itself surrounded by a light gray background. Several mentions are made in the superconductor episode of publishing or reading journal articles; these activities are not featured in the medicinal plant episode. And although no real attempt is made to explicate the specific activities or functions of the complex laboratory apparatus that appears in the superconductor labs, the activities of ethnobotany are revealed as mundane: gathering and sorting plant materials, crushing them in a mortar and pestle, systematically searching for pharmaceutical effects, examining and cataloging the samples. (ibid.:19)

Hornig concludes that "hard sciences" like physics are treated with an air of mystery, and the "soft sciences" are treated as involving only mundane activities. She does not mention the social sciences, since these are apparently almost completely ignored by "NOVA" in its quest to educate the public on the nature of science.

This award-winning series on public television presents an episode illustrating not only the **social stratification** between the United States and Japan, and between physical scientists and biological and social scientists, but also between physical scientists and the lay public. The latter stratification is encouraged by the episode's mystification rather than explanation of the scientific method, with its mumbo-jumbo of dancing cubes, complex laboratory tubes, piles of books and papers, blackboards filled with equations, references to journal articles, superconductors with "almost magical power," and planes that "could be flown merely by the pilot's own thoughts." What this episode reveals is support for Figure 1-1. The top curve represents the **cultural values** associated with economic development, such as "achievement and success," economic "progress," and "material comfort," illustrated by our race with Japan for economic and technological supremacy. And there are also such cultural values as "democracy" and "equality" implicit in an educational program on public TV at this time in history. Yet the bottom curve reveals a pattern of **social stratification** that opposes those latter values, values that would support genuine education rather than mystification. If we are indeed in a race for the survival of humanity requiring genuine education as well as knowledge from the social sciences, then this award-winning series becomes a deadly diversion from what we must learn for the continuation of the human species.

Let us shift to an analysis of science fiction on prime-time television—granting that "NOVA" also suggests the presentation of fiction albeit in the guise of fact—by looking at two very popular series. Jane Banks and Jonathan David Tankel's (1990) analysis of "Star Trek" and "Star Trek: The Next Generation" sees these shows as invoking a hymn to the wonders of physical and biological science technologies. The mission of the Starship *Enterprise*—a name that conveys links between those technologies and economic development—is "to seek out new life and new civilizations, to boldly go where no man has gone before." The "Prime Directive" supposedly guiding the crew of the Starship *Enterprise* is to avoid interfering with the evolution of life and society encountered anywhere in the universe, suggesting the importance of the **cultural value** of "equality." But opposed to this apparent celebration of the worth of all cultures, the *Enterprise* and its crew succeed in transforming almost every culture they encounter in the direction of industrial culture, implying the importance of **social stratification**. All forms of life are viewed in terms of their progress on the "industrial scale." The most advanced forms of life are defined as those which have carried industrialization furthest, such as the members of the United Federation of Planets and Earth in particular.

Captain Kirk, Spock, McCoy, Chekov, and Uhuru of "Star Trek," as well as Picard, Riker, Worf, Yar, LaForce, and Crusher of "Star Trek: The Next Generation," express humanistic ideals, once again illustrating the **cultural value** of "equality." And the viewer cannot avoid awareness of the multicultural, multiracial, and multispecies composition of these officers. Yet at the same time the hundreds of individuals on the crews of these starships remain faceless and invisible, again implying patterns of **social stratification.** These starships are heavily-armed military vessels organized in a highly bureaucratic fashion, which is a highly stratified mode of social organization. Life-and-death decisions affecting the entire crew along with the officers are continually made by the captain. The invisibility of the crew makes it easier for the viewer to pay little attention to the hierarchical nature of relationships on the starship. There are also subtle division among these officers, which reflect the traditional hierarchies of the twentieth century. The two captains are adult white males from the United States or Western Europe. In "Star Trek: The Next Generation" it is only the junior staff that is composed of women, minorities, and a young white male.

The focus on technology as a beneficent force that will solve most problems, and not as something that can create problems, is carried very far:

Life aboard the *Enterprise* is denatured, rendered uniform and sterile by technology. But when crew members want difference, texture, or grit in their lives, technology can provide that, too, via the holodecks. These are chambers where a person's fantasy can be programmed into a computer, resulting in a three-dimensional, fully functional holographic world. These decks, which are nothing so much as 24th-century theme parks, offer the crew all the exotic ambiance and texture they are likely to get on the starship. (ibid.:33)

The holodecks provide environments that enable the crew to accept without challenge whatever problems they face in their everyday lives. Those environments are essential if indeed there are fundamental contradictions in their lives that they remain unable to resolve, such as that between **cultural values** emphasizing "equality" and patterns of **social stratification** on the *Enterprise*. And the fact that those holodecks have been set up on the *Enterprise*—analogous to our own theme parks—is further evidence for the existence of those contradictions.

RELATIVE DEPRIVATION: A MISSING LINK

It is no accident that attention has been paid to the problem of relating macrosociology to microsociology, for within that relationship lies the basis for understanding social and cultural change. It is change within one situation after another that, over time, yields changes in social structure, just as a vision of a different kind of social structure can guide change within a given situation. It is exactly with this idea in mind that Figure 1-3 was constructed so as to include no less than eight situational concepts. Given our present effort to extend our web analysis of the forces that contribute to the escalating gap between aspirations and fulfillment sketched in Figure 1-1, a focus on the situation can provide a missing link to that analysis. For example, just how do the structures of anomie, alienation, and social stratification in fact come together within any given scene so as to yield fundamental problems for the individual? To illustrate further, Merton's analysis of anomie suggested that crime could result from a situation where the **cultural value** of "achievement and success" is heavily emphasized yet where **social stratification** stands in the way of fulfilling that value. But exactly how do those structures come together within any given scene?

Our focus in this example is not on crime but rather on prejudice, and the situational concept at the center of this illustration is **relative deprivation**, introduced in Chapter 1 and defined in the glossary as "the individual's feeling of unjustified loss or frustration of value fulfillment relative to others who are seen as enjoying greater fulfillment." At issue is the question of whether the literature on relative deprivation meshes with and supports our analysis of increasing anomie and alienation in modern society, as portrayed in Figure 1-1. In particular, do our basic patterns of culture and social organization point us in the direction of increasing prejudice against minority groups despite all efforts to educate people to at least be tolerant of others and at best to learn from them? Do those patterns work through situational forces like relative deprivation? And are those patterns of culture and social organization in turn held in place by a bureaucratic **worldview** and cultural paradigm? If the answers to these questions are affirmative, then our specialized efforts at education with respect to prejudice will prove to be largely fruitless, as they are pitted against the fundamental structures of modern society. Further, we might expect that prejudice against minorities will continue to escalate throughout the modern world.

Our focus is on a classroom experiment that builds on the concept of relative deprivation. The idea emerged from studies of the U.S. soldier during World War II, as illustrated by this analysis of greater feelings of relative deprivation among married as well as older men:

> The drafted married man, and especially the father, was making the same sacrifices as others plus the additional one of leaving his family behind. This was officially recognized by draft boards. . . . Reluctance of married men to leave their families would have been reinforced in many instances by extremely reluctant wives whose pressures on the husband to seek deferment were not always easy to resist. . . . The very fact that draft boards were more liberal with married than with single men provided numerous examples to the drafted married man of others in his shoes who got relatively better breaks than he did. Comparing himself with his unmarried associates in the Army, he could feel that induction demanded greater sacrifice from him than from them; and comparing himself with his married civilian friends he could feel that he had been called on for sacrifices which they were escaping altogether. Hence the married man, on the average, was more likely than others to come into the Army with reluctance and, possibly, a sense of injustice.
>
> Or take age. Compared with younger men—apart now from marital condition—the older man had at least three stronger grounds for feeling relatively greater deprivation. One had to do with his job—he was likely to be giving up more than, say, a boy just out of high school. Until the defense boom started wheels turning, many man in their late twenties and early thirties had never known steady employment at high wages. Just as they began to taste the joys of a fat pay check, the draft caught up with them. Or else they had been struggling and sacrificing over a period of years to build up a business or profession. The war stopped that. Second, the older men, in all probability, had more physical defects on the average than younger men. These defects, though not severe enough to satisfy the draft board or induction sta-

tion doctors that they justified deferment, nevertheless could provide a good rationalization for the soldier trying to defend his sense of injustice about being drafted. Both of these factors, job and health, would be aggravated in that a larger proportion of older men than of younger men got deferment in the draft on these grounds—thus providing the older soldiers, like the married soldiers, with ready-made examples of men with comparable backgrounds who were experiencing less deprivation. Third, on the average, older men— particularly those over thirty—would be more likely than youngsters to have a dependent or semi-dependent father or mother—and if, in spite of this fact, the man was drafted he had further grounds for a sense of injustice. (Stouffer et al. 1949:125–26)

The concept of relative deprivation derives its utility by drawing on both social organization and culture and bringing those forces into the momentary situation. First, there is reference to a pattern of social hierarchy or **social stratification** behind the "relative" idea within the concept: married men compare themselves with unmarried draftees and also with married men who obtained deferment, and older men compare themselves with younger draftees as well as older men who were deferred. But this is not a simple hierarchy based on marital status or age alone. Rather, it is feelings about "deprivation" from the fulfillment of **cultural values,** which is the basis for the emerging hierarchies. For example, older men came to feel deprived of opportunities for fulfilling work-related cultural values like "material comfort" just when they were beginning to enjoy a degree of economic success. And people-oriented cultural values like "equality" were also at stake among older men with more health problems than younger men. As a result, what we have here is an illustration that parallels the gap between cultural aspirations and opportunities for fulfilling them that was presented in Figure 1-1. It is not stratification alone that yields the gap but rather stratification coupled with aspirations derived from cultural values, with stratification standing in the way of fulfilling those values.

But there is more to the story of **relative deprivation** that suggests its importance. It also helps us to understand in general how structures from culture and social organization come to have an impact within a momentary scene. Other evidence of the situational impact of relative deprivation comes from studies of revolutions and civil disturbances, to be discussed in Chapter 4 (see, for example, Gurr 1968). There is controversy within the literature of sociology over how far this one concept can be carried to explain collective protests (Sayles 1984), but there is some agreement that relative deprivation is an important factor that is involved. The key problem in this lack of agreement seems to be a desire to predict the occurrence of such disturbances, and any one factor such as relative deprivation is limited in just how far it can take us. However, if we move away from an effort to predict and toward an effort to understand—which is closer to our web approach to the scientific method—then this concept emerges as a very important one. Indeed, it appears to constitute a missing situational link between patterns within social structure that yield the gap between aspirations and fulfillment in Figure 1-1 and patterns within the structure of the individual such as alienation and addiction. It helps us to understand both the situation and the structures producing it.

To learn more about how relative deprivation can yield insight into complex situations we turn to Jack Levin's (1968) classroom experiment, which was the basis for his doctoral dissertation in sociology (see also Levin 1975; Phillips 1979:185–88). Levin attempted to probe the causes of prejudice by looking to the individual's "social frame of reference" or, in our terms, the individual's **worldview.** We might assume, following the arguments in previous chapters, that we all employ a bureaucratic worldview. But some of us might emphasize that worldview more than others, who might employ a scientific or interactive worldview to a very limited extent. If this difference in emphasis can indeed be measured, then we might examine what the repercussions of that difference on the individual's behavior might be. More specifically, would a bureaucratic or outer-oriented *Weltanschauung* tend to be associated with feelings of injustice, deprivation, or a gap between aspiration and fulfillment? And, following the frustration-aggression theory that has developed within psychology around Freudian ideas, would such feelings in turn come to be translated into hierarchical behavior, such as acts of verbal aggression or prejudice? By contrast, would a scientific or interactive worldview deter such feelings of injustice and also deter feelings of prejudice?

Levin devised a procedure for measuring differences in **worldview** (or in his terms "frame of reference"), and he incorporated it within his study of 124 undergraduates from Boston University's School of Public Communication. He administered a questionnaire with a series of paragraphs like this one:

> Mary was in her freshman year of high school and had just received a B on her first algebra examination. Joan, who was Mary's best friend, got an A on the same examination. The class average for the exam was C. Last year in junior high school, Mary had received a C in mathematics. This year, Mary's highest exam grade was an A in French
>
> Without referring back to the paragraph on the preceding page [where the above paragraph was located) what do you think is the most accurate way to describe Mary's grade on her first algebra examination? (check only one)
>
> 1. _____ Mary's algebra grade was higher than the class average
> 2. _____ Mary's algebra grade was lower than her best friend's grade
> 3. _____ Mary's algebra grade was higher than her last year's grade in mathematics
> 4. _____ Mary's algebra grade was lower than her grade in French

The questionnaire was designed to separate those who checked either response 1 or 2, comparing themselves with others, from those who checked 3 or 4, comparing themselves with their own performances. The responses of the other-oriented group suggests a core aspect of the concept of **relative deprivation:** comparing oneself with others. Those responses also correspond roughly to our own concept of bureaucratic **worldview,** with its emphasis on an outer orientation as illustrated by attention to social stratification, bureaucracy, and cultural conformity. The answers of the self-oriented group, by contrast, point away from social comparison and toward the individual's learning about his or her own patterns of behavior. This corresponds roughly to our concept of scientific or interactive worldview, which involves a systematic learning process based on seeing new knowledge in relation to previous knowledge.

Levin's concrete procedure gives us further insight into the nature of these two worldviews. Metaphorically, we might think of them as suggesting a seesaw and a stairway with very wide steps, respectively. The up-or-down position of an individual on a seesaw suggests the stratification of the bureaucratic worldview, involving a context of scarcity where one person's achievements are at the expense of another's. The stairway, by contrast, is not a situation of scarcity. Providing that the stairs are very wide, one can learn to climb higher and higher without threatening others, based on one's own experiences. As for the results Levin obtained in this part of his experiment, all students tended to compare themselves with others in a series of paragraphs like the one quoted above, suggesting an orientation to a bureaucratic **worldview.** However, some students were somewhat less oriented in this way, suggesting at least some orientation to a scientific or interactive worldview. And it was this difference among the students, limited as it was, that enabled Levin to gain a tentative idea about the impact of each of these orientations on the individual's behavior. More specifically, Levin was interested in whether or not an outer or relative orientation, when coupled with deprivation so as to yield relative deprivation, would translate into prejudice against a minority group.

In order to add a component of deprivation to his experiment, Levin did what would now no longer be permitted in social research: he involved the students in what would be for most of them an extremely frustrating experience. Although he attempted to do away with those frustrations at the close of the experiment, modern research procedures generally avoid such experiments because of the possibility of doing irreversible psychological damage to any student with severe emotional problems. Students were led to believe that they were taking an aptitude test for graduate school. They were given twelve minutes to complete a test that consisted of 150 items and were informed that less than 120 correct answers would result in automatic failure. They were also informed that "in similar groups of undergraduates at Boston College and Syracuse University, the average student was able to correctly complete 143 of the 150 items." As if the length of the test along with the small amount of time allowed were not enough to guarantee feelings of failure, fully 50 of the 150 vocabulary-test items were constructed from nonsense syllables. A final touch for creating conditions guaranteeing a sense of failure was a penalty for guessing: each incorrect answer would yield a loss of credit for two correct answers. After the experiment was over, Levin learned that the students almost universally believed in the test's credibility.

Given the "deprivation" students experienced, Levin wished to study the impact of that deprivation on the acts of prejudice employed by those students who were outer-oriented by comparison with students who were somewhat less outer-oriented. In our own terms, would those with more of a bureaucratic **worldview** exhibit a greater degree of prejudice upon experiencing deprivation or frustration than those with the beginnings of a scientific worldview? Levin constructed a before-and-after experiment. Some time prior to his giving the supposed aptitude test for graduate school Levin had administered a questionnaire giving him data assessing prejudice against Puerto Ricans, the "before" part of the experiment. A week later, immediately after the bogus aptitude test, Levin gave them the identical questionnaire, the "after" part of the experiment. By comparing the "after" with the "before" questionnaires, Levin was able to measure any changes in prejudice that followed a frustrating experience, following the frustration-aggression hypothesis linked to Freud's theories. Levin used "the semantic differential," a questionnaire for measuring the meaning an individual attaches to a phenomenon (Osgood et al. 1967):

Directions: Place an "X" in one position between the adjectives of each scale (e.g., _____ : _____ : _____) to indicate how well these adjectives apply in general to Puerto Ricans. Your evaluation should reflect what you believe *many* of the members of this particular group tend to be (what the average Puerto Rican is like), and not necessarily what 100% of them are.

Puerto Ricans

reputable _____ : _____ : _____ : _____ : _____ : _____ : _____ disreputable
knowledgeable _____ : _____ : _____ : _____ : _____ : _____ : _____ ignorant
intelligent _____ : _____ : _____ : _____ : _____ : _____ : _____ stupid
industrious _____ : _____ : _____ : _____ : _____ : _____ : _____ lazy
kind _____ : _____ : _____ : _____ : _____ : _____ : _____ cruel
clean _____ : _____ : _____ : _____ : _____ : _____ : _____ dirty
straightforward _____ : _____ : _____ : _____ : _____ : _____ : _____ sly
reliable _____ : _____ : _____ : _____ : _____ : _____ : _____ unreliable

In comparing results obtained from his "after" measurement of prejudice with those derived from his "before" measurement, Levin found that those individuals classified as relative evaluators (in our terms, those with a bureaucratic **worldview**) tended to increase their levels of prejudice against Puerto Ricans after their frustrating or deprivational experience. By contrast, "self evaluators" (in our terms, those with the beginnings of a scientific **worldview**) did not increase their levels of prejudice in the "after" measurement. Metaphorically, those oriented to a bureaucratic worldview acted like individuals on a seesaw, reacting to frustration by pushing down Puerto Ricans with increased prejudice. In so doing, they would at least rise above Puerto Ricans even if their self-esteem had suffered as a result of failing the test. By contrast, those with an incipient scientific or interactive worldview were like people climbing a stairway with very wide steps. They were primarily interested in improving their own performance by learning to climb ever higher, and a negative reaction to Puerto Ricans would not help them to achieve this goal. The steps are wide, and Puerto Ricans can advance without threatening them, by contrast with the seesaw situation. If they did poorly, then learning why this happened becomes the key to their further advancement, with their attitudes toward Puerto Ricans remaining irrelevant to this goal.

What Levin succeeded in doing was creating a microcosm of the **social** and **personality structures** within modern society, where relative deprivation became the situational link that expressed the impact of social structures on the structures of the individual at a given point in time. Indeed, from this perspective we might see relative deprivation as a missing link that can help us to understand just how the macrocosm affects the microcosm, and vice versa. He created the same kinds of frustrations to be found within modern society's patterns of cultural aspiration, invoking the **cultural value** of achievement and success and systems of **social stratification.** Yet beyond these social structures Levin also brought in structures within the individual that contrasted those with a bureaucratic **worldview** and those with an incipient scientific **worldview**. All of this implies that the very structures of modern society create the basis not only for frustration and anxiety but also for prejudice or verbal aggression. And if frustration and anxiety increase, as suggested by the accelerating gap depicted in Figure 1-1, then it is arguable that so will **relative deprivation** and prejudice. Further, we might remember here Merton's analysis of those same forces from social structure as yielding what might be generally conceived of as yet another type of aggression: crime.

Although Levin emphasized the genesis of prejudice, our own interest is in the range of problems that would support the diagram of an increasing gap between aspirations and fulfillment sketched in Figure 1-1. Durkheim's analysis suggested links between that figure and suicide, and Marx's analysis suggested links between that figure and alienation. Merton suggested that such a gap might be linked to "retreatism," a species of deviant behavior involving "the rejection of cultural goals and institutional means." For Merton, retreatists include "psychotics, autists, pariahs, outcasts, vagrants, vagabonds, tramps, chronic drunkards and drug addicts" (1949: 142). We might add here the fact that experiments like Levin's where the individual is severely frustrated are no longer allowed for the protection of those individuals. This suggests the dangers such experiments might pose with respect to the mental status of the individual, perhaps even pointing in the direction of suicide. Yet it is ironic to note that the dangers to the mental health of the individual posed by the fundamental structures of society—which appear to yield a situation for the individual similar to the one Levin set up experimentally—remain fully in force.

What this chapter as a whole reveals are many findings from the literatures of sociology pointing to the existence of an accelerating gap between aspirations and fulfillment within modern society. Those findings have to do not with minor factors but rather with the fundamental social structure—both **culture** and patterns of **social organization**—within our contemporary world. These literatures also argue for the existence of that accelerating gap within the **personality structures** of individuals within modern society. We have been able to trace a crucial process—involving **relative deprivation**—by which these social and personality structures exert their influence within the momentary scene. If we were to compare this result with present arguments for the existence or lack of existence of this gap, we would note that the weight of sociological knowledge that the foregoing analysis invokes exceeds by far the weight invoked by traditional specialized analyses. This suggests not the certainty of the foregoing analysis but rather its credibility, granting the general difficulty of assigning credibility to the social sciences by comparison with the physical and biological sciences. Given that credibility, we are left with a highly probable and threatening conclusion: that basic problems within modern society such as **anomie** and **alienation** are accelerating.

In order to encompass a problem as broad as this one it was essential to move beyond our traditional highly specialized interpretation of the scientific method, which results in little communication among those within sociology's forty sections of the American Sociological Association. Instead, we employed an alternative interpretation of the scientific method, the web approach discussed in Chapters 1 and 2, and this has yielded a network of findings that directly and indirectly supports the schematic diagram in Figure 1-1. This suggests the utility of this methodological and theoretical approach of integrating knowledge within the discipline around any research problem and not only those that are fundamental to social and personality structures and that involve long-term change. This approach requires our defining sociological concepts so abstractly that they are broad enough to encompass the range of data and problems within the discipline. Our choice is not between employing general yet vague and eclectic concepts and employing middle-range concepts that are closer to our data. Rather, it is between employing abstract concepts that are broad enough to encompass our middle-range concepts and avoiding abstract concepts so as to continue to yield what we presently have: a shattered discipline.

One insight emerging from the foregoing analysis is the relative invisibility of the fundamental forces at work within modern society, such as **anomie, alienation, social stratification, cultural values, personality structure,** and **worldview.** Many would view the importance of such forces as too abstract or vague to be given serious attention, preferring instead more visible phenomena like poverty, crime, war, and suicide. Yet if we continue to fail to make such relatively invisible phenomena visible, then it appears that we will remain unable to understand and confront basic problems that will continue to escalate and finally explode. Another insight is the importance of the concept of **relative deprivation** for understanding just how social and personality structures have an impact on any given momentary scene. Much of recent work in sociology has centered on the importance of the momentary situation for any deep understanding of how social change occurs as well as the nature of social problems, and this abstract concept helps us in this area. It helps us to understand our abstract sociological concepts as not merely reified entities that can be safely ignored within any concrete analysis but rather as powerful forces that have repercussions from one moment to the next in every scene.

Granting the depth of modern problems, they are not rooted in human biology but rather stem from forces that can be altered. Figure 1-5 sketched a path toward such change, one that encompasses shifts in our approach to the scientific method in sociology, our **worldview,** our **cultural values** and patterns of **social stratification,** our **anomie** and **alienation,** and our patterns of **relative deprivation.** That sketch points toward a possible future where the fulfillment of aspirations continues to increase, reducing the gap between what we want and are able to obtain. Just as in the case of the multiple sources of evidence that support the escalation of that gap if those forces remain invisible and unchallenged, so are there multiple sources of evidence supporting Figure 1-5. These will be presented in Chapter 4. There we shall see that this possibility works, metaphorically, much like a pendulum, where movement in the direction of solutions depends on momentum derived from movement in the opposite direction: awareness and understanding of the problem. And we shall also see that momentum obtained from movement in the direction of solutions can yield, in turn, further momentum in the opposite direction: awareness and understanding of our problems.

4

The Web Approach Illustrated: Addressing the Invisible Crisis

To elaborate on our own present situation in contemporary society, it appears to be much like that of the faceless crew of the Starship Enterprise, as described in Chapter 3, whose off-duty time is often spent within the fantasy world of the holodecks. We have learned to release some of the tensions embedded in our contradictory lives within a world of passive entertainment and consumption, and in this way we fail to challenge the forces producing those tensions. Or for a more sociological illustration of our situation we might turn to Vidich and Bensman's (1960) analysis of "Springdale," a small town in New York State, where large gaps developed between the Springdalers' aspirations for success, friendship, and self-determination and what they actually achieved:

> The technique of particularization is one of the most pervasive ways of avoiding reality. It operates to make possible not only the failure to recognize dependence but also the avoidance of the realities of social class and inequalities. The Springdaler is able to maintain his equalitarian ideology because he avoids generalizing about class differences. . . . Thus a new purchase is talked about only in terms of the individual who makes it, rather than the class style of the purchase. . . . The realization of lack of fulfillment of aspiration and ambition might pose an unsolvable personal problem if the falsification of memory did not occur, and if the hopes and ambitions of a past decade or two remained salient in the present perspective. . . . As a consequence, his present self, instead of entertaining the youthful dream of a 500-acre farm, entertains the plan to buy a home freezer by the fall. . . . The greatest dangers to a system of illusions which is threatened by an uncompromising reality are introspection and thought. . . . The major technique of self-avoidance is work. The farmer and the businessman drive themselves in their work almost to the point of exhaustion. (pp. 299, 303, 311]

It is now almost half a century after the study of Springdale, and much has changed markedly. Yet at the same time there remain fundamental similarities between the Springdalers and the inhabitants of modern society. Our own bureaucratic approach to the scientific method apparently teaches us sociologists "the technique of particularization" by emphasizing research at a low or middle-range level of abstraction. That same approach also teaches us "the falsification of memory" by emphasizing a focus on the present rather than long-term aspirations such as depicted within Figure 1-1. And that approach teaches us as well to bury ourselves in specialized work and thus avoid thinking about the failures of our own aspirations as sociologists. Just as in the case of the Springdalers, this behavior manages to give us at least some measure of satisfaction:

> Because they do not recognize their defeat, they are not defeated. The compromises, the self-deception and the self-avoidance are mechanisms which work; for, in operating on the basis of contradictory, illogical and conflicting assumptions, they are able to cope in their day-to-day lives with their immediate problems in a way that permits some degree of satisfaction, recognition and achievement. (ibid.:320)

Our bureaucratic approach to the scientific method backed up by our **bureaucratic worldview** enables us, like the Springdalers, to achieve at least some satisfaction, where further achievement "does not appear to lie within the framework of . . . [our] **social structure**" (ibid.).

Yet social structure is not biologically given: it can be changed. In this chapter we emphasize the promise of sociology for changing both the **social structure** of our discipline as well as that of society at this time in history. It is a promise that can help us to define the twenty-first century not as the age of death, despair, and destruction but as the age of the social sciences. Granting the existence of widespread pessimism and cynicism among us, feelings that have pushed us into hiding within our specialized areas of knowledge, our beliefs in the promise of sociology have not yet died. Those beliefs lie deeply buried under much of the history of a twentieth century that has disappointed not only sociologists but modern society as a whole, for the Enlightenment dream was never limited to academicians. Fortunately, we appear to be at a stage in history where the credibility of traditional interpretations of the scientific method—along with the social sciences in general—are undergoing reexamination. Given this window of opportunity, we believe that what is required most of all is an approach to the scientific method that yields increasingly credible knowledge that can serve as a platform for addressing the fundamental problems of modern society. We have no more than begun to illustrate this approach to the scientific method, which competes with procedures used within sociology for well over a century. Yet we believe it carries with it the potential for reviving the promise of sociology and the Enlightenment dream for modern society.

In this chapter we shall sketch directions for changing sociologists' paradigm and **worldview**—as well as the worldview of social scientists, academicians, technologists, and people throughout modern society—from a bureaucratic to an interactive or scientific one. The sketches will be based on using the same kinds of interactive feedback loops used in Figures 1-4 and 1-5 and will necessarily be quite brief, given our commitment in this chapter to provide some illustrations of sociological knowledge on how we might confront the invisible crisis of modern society. Yet we shall have an opportunity in Chapters 5 and 6, where we examine the implications of our analysis for sociology and society, respectively, to fill out those sketches. Our focus there will not be on some distant utopia, where we are beyond the urgent problems that modern society faces at this moment. Rather, we shall emphasize our present situation, but with two additions that have been sketched in the foregoing chapters. One consists of ideas backed up by broad sociological knowledge as to the problems within our present understanding of the scientific method as well as within our worldview, and how those problems in turn contribute to our invisible and visible crises. Another, also bolstered by that broad knowledge and emphasized in this chapter, consists of ideas on how we might confront those crises ever more effectively, helping to create an age of social science.

We begin by going back to the future, returning to ideas within Chapters 1, 2, and 3 in order to build on their implications for an interactive—versus bureaucratic—approach to the scientific method within sociology, as well as for understanding and confronting basic problems within modern society. We do this in our initial section, "Interaction and the Crisis: General," where the idea of interaction—as illustrated by concepts like **social interaction** and **interactive worldview**—becomes a focus for pulling together those earlier ideas and moving beyond them. Whereas in this section our emphasis is on a high level of abstraction, in our second and final section we emphasize a range of concrete phenomena. The idea of interaction and the problem of the invisible crisis within modern society are still central to our orientation in this section, "Interaction and the Crisis: Specific." And, of course, we continue to employ our abstract concepts as our basic tools for analysis. We make use of studies of revolution, look to one example of a revolution, and also take up some studies of the individual in relation to emotional awareness. By placing these studies within the context of our web approach to the scientific method as well as an interactive worldview, they can add to our understanding both of the nature of our crisis and how we might confront it. And they give us the basis for a more thorough treatment of that confrontation in Chapters 5 and 6.

INTERACTION AND THE CRISIS: GENERAL

If Chapter 3 centered on four concepts—anomie, alienation, social stratification, and relative deprivation—in order to address the problem of the escalating gap between aspirations and fulfillment in modern society, in this section we give special emphasis to one concept to address that same problem: interaction. It is an idea that is, of course, absolutely central to the discipline of sociology, yet paradoxically our specialized **worldview** has taught us to miss out on that centrality. We are still committed to using twenty-five other concepts in order to uncover the complexity of human behavior. We shall begin this section by going very far back to the future, even to the nature of physical and biological structures as well as the process of biological evolution, for it is there that we must begin to learn about the nature of interaction. We then return to some of Nietzsche's, Dewey's, Kuhn's, and Kincaid's ideas, as spelled out in Chapter 1, and consider their implications for the idea of interaction as well as for an approach to changing **social structure**. We then proceed to examine our twenty-six concepts with new eyes, this time contrasting more sharply a bureaucratic with an interactive worldview. Just what are we accomplishing with this system of concepts, and just how are we accomplishing it? Finally, we present one figure with feedback loops in an effort to carry further ideas in Figures 1-4 and 1-5. It helps us to develop a more realistic view than Figure 1-4's view of the invisible crisis of modern society.

The Idea of Interaction

It was in Chapter 2's first section, "Cultural Paradigms and Worldviews," that we journeyed far backward into evolutionary history and then far forward to the ideas of Nietzsche, Dewey, Kuhn, and Kincaid. Implicit and to some extent explicit throughout that material was the contrast between a relative lack of interaction—illustrated by social stratification and bureaucracy—and interaction. There we noted our definition of **physical structures** as involving at least some degree of interaction. And we also saw the contrast between **physical structures** and **biological structures,** where the latter were defined as interacting to a relatively great extent. This was illustrated in the case of biological evolution, based on the long-term interaction between organisms and their environments. At this time, however, let us not pass over these facts so lightly. For if the physical and biological sciences have gained a great deal of well-deserved credibility for their achievements, our own usage of the idea of interaction along with the concepts of physical and biological structures—embedded as it is very deeply within our own web approach to the scientific method—

should strengthen considerably the credibility of that usage along with the achievements of our discipline, for it is linked closely to interaction. And more than credibility is at stake here, for we can also gain further insight into the invisible crisis of modern society to the extent that we learn to change from a less to a more interactive approach to research.

What Nietzsche does for us with his emphasis on the death of God is to teach us about the death of an autocratic culture where the church or the state rules the individual, by contrast with one where the individual is free to develop as a human being. Simmel, who was influenced by Nietzsche, wrote about something quite similar when he contrasted "objective culture" with "subjective culture." In our own terms we can think about a transition from a society emphasizing **cultural conformity, social stratification, labeling, relative deprivation, bureaucracy,** and a **bureaucratic worldview** to one emphasizing **social interaction** and an **interactive worldview.** But we should be careful not to think completely in either-or terms, where a bureaucratic worldview is bad and an interactive one is good, for this teaches us to cut away the very ground on which we stand. Relative to preindustrial societies, our bureaucratic modern society has achieved a great deal in most areas of life, not the least of which are the very cultural values espoused by Nietzsche and Simmel. Since we Springdalers must live with that worldview at least for the present, let us give thanks to a way of life that "permits some degree of satisfaction, recognition and achievement," so long as we realize that our window of opportunity for confronting our basic problems and changing that worldview is closing rapidly. We cannot destroy our home until we have at least some idea of how to construct another.

John Dewey's recognition of the importance of the physical and biological sciences in shaping modern society, and his recognition of our failure to achieve a credible and effective social science, were complemented by a vision of an educational society—doing justice to Nietzsche and Simmel— centered around the continuing development of every single individual:

> Democracy has many meanings, but if it has a moral meaning, it is found in resolving that the supreme test of all political institutions and industrial arrangements shall be the contribution they make to the all-around growth of every member of society. ([1920]1948:86)

The physical and biological sciences were so effective largely because they replaced an authoritarian or stratified approach to knowledge with an interactive one, with the ideal that the findings of any scientist could overturn centuries of belief by the most noted of authorities. And it is exactly this interactive approach to the scientific method that is largely lacking within our present efforts to develop a science of sociology. Dewey's vision of democracy is interactive as well, pointing the individual away from bowing down to authorities and conforming to untested norms throughout society. Rather, he envisaged a learning or scientific society where the individual is free to test all ideas on the basis of experience.

Although Thomas Kuhn centered on the history of the physical and biological sciences, we can extend his analysis much further so as to include not only the social sciences but also the general process of social change, since we can use our web approach to bolster his ideas with the literature of the social sciences. Here, our pendulum metaphor—which is based on the idea of interaction—can help us to understand how to address the invisible crisis of modern society. We Springdalers require at least some way of dealing with our research problems even if it admittedly does not swing our pendulum very far in defining problems of a magnitude that is in accordance with the promise of sociology. However, if we can manage to develop an approach to the scientific method that—because of its interactive nature—enables us to swing that pendulum further so as to define problems like the invisible crisis, then that will also give us momentum to swing further in the opposite direction. That latter direction has to do with solutions for both substantive and applied problems, and—just as in the case of the physical and biological sciences—it is central to the development of credibility and insight for sociology. Kuhn helped us to understand the importance of developing a new scientific paradigm if we are to abandon the old one, granting the many forces that stand in the way of such change. Equally, we require a new **worldview** and cultural paradigm to have a chance of changing the old ones.

Harold Kincaid takes us from these general Kuhnian ideas to more specific ones: a web approach to the scientific method within the social sciences. It is an approach in accord with the interactive ideals of the scientific method, where knowledge is not chopped into watertight compartments where sociologists are to be found communicating only with other sociologists within their own specialties. And this orientation is open to indirect evidence, for there is no bowing down to the supposed accuracy of complex and mathematical tools of measurement, analysis, and prediction. Yet he leaves open the question of just how this approach might be applied within sociology, especially to problems as fundamental as the invisible crisis of modern society. How abstractly should we define our concepts? To what extent must we adhere to the original definitions? To what extent must we invoke the entire theory of a given sociologist and not just one or two concepts? How are these concepts to be linked to one another? How can we proceed to test their utility? If we are indeed to attempt to alter our approach to research within sociology, how do we take into account existing procedures along with the many barriers standing in the way of changing a scientific paradigm? Perhaps most important, what do we do about the cultural paradigm and **worldview** that works to support our traditional scientific paradigm? How can a new scientific and cultural paradigm resolve the contradictions within the old ones?

Interaction among a Web of Sociological Concepts

Chapters 1 and 2 spell out a general methodological approach for sociology, and Chapter 3 applies that approach to penetrate the nature of the invisible crisis of modern society. It is in those chapters that preliminary answers to these questions are put forward. At this point we are in a position to clarify those answers by centering on this fundamental problem of an accelerating gap between aspirations and fulfillment and by contrasting the interactive nature of bureaucratic and interactive research procedures. Bureaucratic research procedures have taught us a great deal, just as our **bureaucratic worldview** has accomplished a great deal for modern society. Paraphrasing Marx's analysis of capitalism, never before in human history has our knowledge of human behavior developed so far and advanced so rapidly, giving us an absolutely unprecedented wealth of knowledge. Yet at the same time modern problems have escalated to unprecedented magnitudes, just as Marx saw an escalating contradiction between the forces of production and the relations of production. Given the problematic situation of modern society at this time in history—analyzed throughout Chapter 3 and not simply within Marx's theory—it is no longer sufficient for us sociologists to conduct business as usual and avoid the huge problems we face. We must somehow construct the scientific tools that are required to address those problems. Here, we shall proceed to contrast the interactive nature of the traditional and the web approach, centering on the twenty-six concepts within Figure 1-3.

Although that figure was used primarily to illustrate only one aspect of our web approach to the scientific method—defining concepts at a high level of abstraction—it can also help us to understand that web approach in general. With respect to the definition of a problem for investigation, our traditional or bureaucratic approach steers us away from the kinds of fundamental problems suggested by Figure 1-1, which encompasses both **culture** and **social organization** and has to do with very long-term change. That figure includes those concepts as well as much more, pointing us toward examining the interaction among all of these concepts within any given scene. That figure's call for defining concepts at a high level of abstraction contrasts with the low or middle-range level within traditional approaches, which results in the isolation of studies from one another. By contrast, greater abstraction builds bridges among specialized fields, just as a focus on **social stratification** can link studies of racism, sexism, ageism, classism, and ethnocentrism, where those concepts are defined at a lower level of abstraction. In this way we can take into account both interaction between social stratification and concepts at lower levels as well as the integration of knowledge from studies of racism and sexism, or among lower levels of analysis. Finally, our use of boldface for these concepts in Chapter 3 suggests a reflexive approach to research, where we attempt to become conscious of our limited approach to using language from one moment to the next.

In addition to looking at Figure 1-3 as a whole, we can probe that figure more specifically to uncover further contrasts between a bureaucratic and an interactive approach to the scientific method. Let us start by moving downward from the top row to the middle row to the bottom row. Traditionally we sociologists are divided in a great many ways, such as the walls between those emphasizing **social structure,** those emphasizing the situation, and those emphasizing the **individual.** The diagram implicitly calls on sociologists to trace the interactions among these three areas, given the assumption that all of these concepts can be employed simultaneously to help explain the complexity of any situation whatsoever. For example, there is the link between **social stratification** and **cultural values** within the top row and **relative deprivation** in the middle row, as illustrated by Levin's classroom experiment in Chapter 3. There he found most students to focus on comparing their achievements to one another in a hierarchical way, following an orientation to social stratification. Cultural values were involved in the experiment's emphasis on "achievement and success." And as a result, under conditions of stress these students tended toward feelings of **relative deprivation**—comparing their achievements with those of others—and manifested their feelings of deprivation relative to others by increasing their prejudice or **labeling** with respect to Puerto Ricans.

As another example, illustrated by the analysis of "The Dr. Ruth Show" in Chapter 3, we can see **anomie, alienation,** and **worldview** as linked to **conformity.** Here, Dr. Ruth takes advantage of basic problems experienced in modern society to gain popularity for her television show and exert influence on her audience. Following the analyses of anomie and alienation in that chapter, we can see those phenomena as increasing as a result of the increasing gap between aspirations and fulfillment portrayed in Figure 1-1. And all of this appears to be linked to a bureaucratic worldview. As a result of these problems being experienced by the individual, he or she learns to repress or hide them from self and others in order to obtain some satisfaction from day-to-day living, just as the Springdalers succeeded in doing. Dr. Ruth appeared to offer the possibility of uncovering those feelings, particularly within the area of sexual relations, with the possibility of closing that gap between aspirations and fulfillment. Instead, Dr. Ruth actually offered dependence on her expert opinions, robbing her viewers of their autonomy within this intimate emotional area by influencing them—in their hour of need—to **conform** to her own philosophy. And the result of such conformity would be yet further **alienation,** further **anomie** in society as a whole, and further support for the **bureaucratic worldview,** which holds all of these problems in place and leads to their escalation.

It is also important to examine interaction among the rows of Figure 1-3 to develop a deeper understanding of social and cultural change. The foregoing examples invoked links between structures—whether social or individual—and the momentary scene, or between the top and bottom rows and the central row. Whereas structures emphasize uniformities among long sequences of scenes, situational concepts have to do with a momentary scene. When we fail to link structural with situational concepts, which characterizes a failure within much of sociology over most of its history, we fail to learn very much about how social and cultural change actually occur. It is within a scene that a change is initiated, and when that change is repeated in scene after scene we are dealing with a structure. In presenting Weber's analysis of *The Protestant Ethic* in Chapter 3, we used all of our situational concepts—**definition of the situation, label, relative deprivation, reinforcement, conformity, deviance,** and **social interaction**—to characterize Weber's illustrative material. Yet although he was conscious of the importance of such scenic material within his examples, he failed to develop the situational concepts that modern sociology has produced to analyze the complexity of *any* scene whatsoever. By using those concepts as well as structural ones we are able to achieve a more dynamic analysis, going beyond the presentation of good illustration to learn about the origins of structures.

In addition to the interaction among the rows within Figure 1-3 as a way of understanding more fully the basic problems in modern society, we can look to the interaction among the columns, such as those headed by culture and social organization. To illustrate, the invisible gap between aspirations and fulfillment portrayed in Figure 1-1 will remain invisible to the extent that we fail to link **cultural values** with patterns of social organization like **social stratification** and **bureaucracy,** patterns that tend to limit the fulfillment of those aspirations. That failure is partly due to a commonly used definition of social structure solely in terms of patterns of social organization, excluding patterns of culture from that definition. In that way, culture no longer comes to be taken seriously as a structural phenomenon, and it takes a back seat within structural analyses. By contrast, we can proceed to define social structure so as to include both **culture** and **social organization,** granting that this departs from the usage of many sociologists. By so doing we are able to pay serious attention to culture as an important structure, and we can elaborate on the nature of that structure by examining the range of **cultural values.** By so doing we can become sensitive to the increasing **anomie** with which we moderns are confronted, as illustrated within Figure 1-1: the growing gap between aspirations or cultural values and their fulfillment, as deterred by patterns of social stratification and bureaucracy.

Another example of interaction between the columns of Figure 1-3 has to do with the Levin experiment presented in Chapter 3. There we noted that the **bureaucratic worldview** of most students, with its orientation to compare themselves with others, led to **negative reinforcement** in relation to the **cultural value** of "achievement and success" when they learned of their poor performance relative to others. As a result they suffered feelings of **relative deprivation,** and that in turn influenced them to **label** Puerto Ricans more negatively than previously. Of course, a fuller analysis would require us to move beyond these concepts so as to take others into account. For example, that bureaucratic worldview is in turn linked to patterns of **social stratification** and **bureaucracy** as well as to patterns of **cultural conformity.** It is also linked to the phenomena of **anomie** and **alienation.** And we might also add the little-used concept of **addiction,** defined abstractly as "the individual's subordination of individuality to dependence on external phenomena" and not narrowly as physiological dependence on a substance. That concept helps us to understand why those students tended to look outward and passively compare themselves with others rather than actively compare themselves with their own previous performance. Following Simmel, they face the enormous problem of defending their individuality within the modern world.

Yet another illustration of interaction within our approach has to do with our use of boldface on occasion when using the concepts in Figure 1-3. Following the rationale presented in Chapter 3 this can serve as a reminder of the importance of the full range of concepts within that figure and not merely the concept at hand. Our bureaucratic worldview constrains us to seek simplicity in an either-or fashion, avoiding the complexity of any given scene. Every word we use divides all phenomena into whatever fits within the category that word calls for and whatever is left outside that category. Yet the world is in fact far more complicated, and an interactive worldview requires that we learn to think in more complex ways. For example, when we see a concept in boldface we can understand its linkage to some twenty-five other concepts. But even more generally, that boldface can remind us of the complexity associated with any word that we use. Here we might recall the efforts by Alfred Korzybski, the founder of general semantics—touched on in the introduction to Part II—to teach people "consciousness of abstracting," where our verbal maps come to be understood as simpler than the territories they supposedly portray. Boldface can, then, help us to see the shortcomings of our own worldview, as illustrated by the very way we have learned to think, speak, and write from one moment to the next, giving us a tool for achieving what Gouldner called "reflexive sociology."

We can carry further our understanding of the interactive perspective within earlier chapters through a closer look at the feedback loops portrayed within Figures 1-4 and 1-5. In Figure 1-4 we examined the interactions among a set of concepts describing forces behind the invisible crisis portrayed in Figure 1-1. And in Figure 1-5 we looked at a set of concepts describing forces behind a possible solution to that crisis. The key differences between those figures is a change from a bureaucratic to an interactive sociological paradigm and worldview. Yet the key question we should be addressing more fully is just how those changes are in fact produced. We will be able to do this to a greater extent by beginning with a more realistic view of our present situation than the completely pessimistic view depicted by Figure 1-4. From there we develop a more realistic view of Figure 1-5, unearthing greater complexity. These diagrams are only schematic, but they are suggestive nevertheless of the complex forces that are creating our invisible crisis as well as those which might resolve it. By turning to feedback loop diagrams instead of relying only on verbal analysis, we are able to increase the systematic nature of that analysis. In this vein we might continue to move further toward models and computer simulation, but by so doing we would be allowing the tail of sophisticated methodology to wag the dog of sociology.

Interaction and Feedback Loops

In Figure 1-4 we presented a causal-loop diagram sketching some of the key forces behind the accelerating gap between aspirations and fulfillment in Figure 1-1. And in Figure 1-5 we depicted a decelerating gap between aspirations and fulfillment and also presented a causal-loop diagram of some key forces behind that deceleration. Together, the two figures also illustrated a general procedure for drawing causal-loop diagrams [see Roberts et al., 1983]. Such a procedure may be new to sociology, yet it holds the promise of moving to a more interactive and sophisticated approach to dealing with the complexities of human behavior. Also, it is a large step in the direction of a full-fledged computer simulation, one that would carry much further our ability to handle such complexity. At this point it would be useful for the reader to review the discussion of those two figures in Chapter 1, for we will now not only revisit them but build upon them. We might think of causal-loop diagrams as a technology for constructing theory, just as procedures like multivariate analysis are technologies for analyzing data. Both kinds of technologies are potentially very useful. In each case there is a danger of becoming one-sided. By emphasizing data at the expense of important theory we run the risk of triviality. Similarly, by emphasizing theory at the expense of data we run the risk of unsupported and frivolous ideas. C. Wright Mills' advice is relevant here: i.e., to shuttle up and down language's ladder of abstraction instead of invoking "grand theory" or "abstracted empiricism."

However, we should bear in mind that sociologists have been presented with many technologies for analyzing data, but hardly any for developing theory. Granting the difficulties involved in learning any new technology, we would find it useful to welcome one that helps us to construct theory, though we should keep in mind the lessons we have learned about the limitations of technologies for analyzing data. This theoretical technology by no means carries with it our whole approach to reconstructing the scientific method. For example, that reconstruction also emphasizes using abstract social-science concepts that are linked within the social-science literature both to less abstract concepts and to one another, thus carrying along with them considerable knowledge from the social sciences. By contrast, causal-loop diagrams have only rarely been used by social scientists, and the concepts that have been employed have generally been common-sense ones. Further, they have frequently been tied to very detailed procedures for computer simulation—procedures involving many largely untested assumptions that have been challenged [see for example Meadows et al., 1972]. Granting that causal loop diagrams do have the potential to obtain greater complexity, it is a potential that should be tied to what we already know in the social sciences if it is to carry that knowledge further.

Figure 4-1 is a causal-loop diagram that repeats the top loop within Figure 1-4 with its focus on society as a whole, eliminating the bottom loop with its focus on sociology. The purpose of this simplification is to clear the way for a more complex and realistic view of that top loop. Let us note that the sole differences between that top loop and Figure 4-1 are (1) the *minus* sign in the center of Figure 4-1, indicating that it is a negative loop by contrast with the *plus* sign in the center of Figure 1-4's top loop, indicating that it is a positive loop, and (2) the *minus* sign outside the lower right side of the perimeter of Figure 4-1's loop, by contrast with the *plus* sign in the same place for Figure 1-4's top loop. All of the loops we have seen in the previous figures presented —1-4 and 1-5—have been positive ones, indicating that all forces involved are moving in the same direction, reinforcing one another and producing acceleration. By contrast, the negative loop in Figure 4-1 involves opposing forces, which constitute a more realistic view of

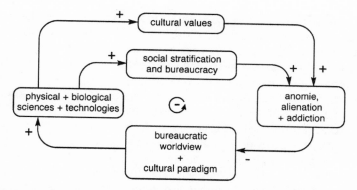

Figure 4-1. Forces within Society Linked to the Accelerating Gap: A More Realistic View.

the actual situation depicted. A causal-loop diagram like Figure 1-4 does not specify the weight of those opposing forces, however, as does a computer simulation, and that weight remains to be assessed on the basis of whatever we have learned about a given situation. In any case, we do not have in Figure 4-1 the continuing acceleration in the same direction that we have in Figure 1-4.

To explain the rationale for this negative loop in Figure 4-1 prior to launching into technical details about how it actually works, in Figure 4-1 we see "anomie, alienation and addiction" as being inversely related to "bureaucratic worldview" and "cultural paradigm," as indicated by the minus sign next to the arrow between the two. If indeed our worldview produces a society with increasing problems such as anomie, alienation and addiction, then it is reasonable to believe that we would question our worldview at least to some extent, rather than blindly follow it and ignore those increasing problems. Further, if there are more visible problems that appear to be increasing, such as the chances for becoming a victim of weapons of mass destruction or the gap between the rich and the poor, those visible problems might also encourage us to question the fundamental assumptions on which modern society rests. Such questioning is suggested by the minus sign, indicating that we do not become ever more committed to the idea that our society is on the right track. Technically, the questioning of our worldview —proceeding further around the loop— yields questioning of all of the other elements of the loop. The plus signs indicate direct relationships, so that a lessening emphasis on our worldview would tend to bring about less emphasis on our biophysical sciences and technologies; less on our cultural values and patterns of social strati-

fication; and in turn less anomie, alienation and addiction. But here the minus sign indicates an inverse relationship or a second reversal. Since the
only way of life we know has managed to reduce those problems, we have
little choice but to emphasize it once again. And so we go around the loop
with alternating reversals supporting and challenging our worldview.

If we now take into account the relative magnitudes of the various forces
in Figure 4-1—based on what we hypothesize from our knowledge and not
anything diagrammed in that loop—then we can understand Figure 4-1 as
not so very different from 1-4. Of course we experience continuing reversals, but we might hypothesize that the emphasis on our worldview, the
development of the biophysical sciences and technologies, our cultural
values and patterns of bureaucracy are more powerful and continue over
a longer period of time than the reverse. And the result would be increasing anomie, alienation and addiction over the long term, although that the
acceleration of those phenomena would not be as great as is depicted in
Figure 1-4. Here we might return to extrapolating Thomas Kuhn's argument about the difficulties of changing a scientific paradigm. Einstein came
up with an alternative theory that resolved Newtonian contradictions, a
theory that was crucial to his ultimately achieving a change away from the
Newtonian paradigm. Without an alternative worldview and cultural paradigm that promise to resolve existing contradictions within society, we
cannot expect a change in paradigms. People cannot be persuaded to cut
out from under themselves the very ground on which they stand, unless
we believe in some alternative ground to which they might move. Figure
4-1 is more realistic than Figure 1-4, but its ultimate outcome is much the
same, because the problems of anomie, alienation and addiction do not accelerate to the same degree.

If we now turn our attention to the top loop of Figure 1-5, the introduction of Figure 4-1 just presented can help us to gain a more realistic understanding of that loop just as it helped us to see Figure 1-4's top loop
more realistically. In Figure 4-1's negative loop there are continuing reversals, but they are based on the fact that there is no alternative worldview
which would gradually build up strength and ultimately supplant present
our bureaucratic worldview and cultural paradigm. However, if indeed an
interactive worldview and cultural paradigm gradually developed, then it
is reasonable to hypothesize that each time there was a questioning of the
worth of our bureaucratic worldview that such an interactive worldview
would gain strength. Over time, then, we might expect the interactive
worldview to replace our bureaucratic one. And this then would yield the
positive top loop we see in Figure 1-5. In this way, we might view our present negative-loop situation as questioning to some extent the adequacy of
our basic way of life, as depicted in Figure 4-1, as a transitional situation,

provided that we are indeed able to construct an alternative worldview and cultural paradigm. Following Kuhn's emphasis on the importance of the presence of an alternative research paradigm if an existing research paradigm is to be replaced, we hypothesize that the same holds true for worldviews and cultural paradigms. Absent such an alternative—one that promises to resolve contradictions within our existing paradigms and worldviews—it appears that the situation depicted in Figure 1-1 will not change, despite the negative loop in Figure 4-1.

The foregoing analysis of interaction in relation to the accelerating gap between aspirations and fulfillment depicted in Figure 1-1 gives us a relatively clear and simple direction for closing that loop: learning to increase various kinds of interaction. There is, first, the general idea of interaction that is so central to an interactive sociological and cultural paradigm. That approach parallels the kinds of interaction we appear to have experienced within the biophysical sciences, and it is illustrated by the solid lines of latitude and longitude in Figure 1-2. Within this general approach, we can look more specifically to the sociological concepts in Figure 1-3. What is required here is to see them as working simultaneously within any given scene and not as isolated from one another with each concept utilized in a different situation. This approach will be simplified in Table 5-1, which points to the importance of taking into account as many as nine categories of phenomena without necessarily using the sociological concepts presented in Figure 1-3. Finally, our interactive approach extends to causal-loop diagrams, which go beyond the one-way analyses so prevalent in the social sciences. Beyond this general discussion, however, if we are to understand that approach more fully, it is essential to examine actual research studies that have adopted an interactive approach. And even beyond such understanding, we must bear in mind the difficulties involved in shifting our research and cultural paradigms, assuming, of course, that there are convincing arguments to do so.

INTERACTION AND THE CRISIS: SPECIFIC

Within our web approach to the scientific method in sociology with its emphasis on the interaction among phenomena, few if any studies can fail to be useful for learning more about our invisible crisis and how it might be confronted. To illustrate our traditional methodology, which points in the opposite direction, here is an analysis of revolutions:

The study of revolutions remains much like the study of earthquakes. When one occurs, scholars try to make sense of the myriad of data that have been collected, and build theories to account for the next one. Gradually, we gain a fuller understanding of them, and the conditions behind them, but the next one that occurs still surprises us. Our knowledge of revolutions, like that of earthquakes, is still limited. (Goldstone 1982:205; quoted in Sztompka 1993:320)

Granting that our knowledge of revolutions is still limited, there is much to learn from research on revolutions outside a narrow focus on attempting to predict their occurrence. Such research can contribute, for example, to our understanding of what are the fundamental problems of modern society, the problems that come to be the basis for revolutions. As another example, such studies can also contribute to our understanding of the impact of efforts to resolve those problems. Such contributions do not yield accurate predictions, yet they can yield important knowledge and useful insights into human behavior in general, provided that we reserve a place for them within our interactive approach to the scientific method.

In this section I do not select studies at random simply to demonstrate the utility of all studies within a web approach, although that could in fact be done. Rather, I select studies that bear directly rather than only indirectly on the problem at hand—how to confront the invisible crisis—and that have come to my attention. Studies of revolution are one example. Revolutions occur in the wake of major problems and not simply minor ones, and our invisible crisis is indeed a fundamental problem. Such studies, then, might help us to understand whether the same forces portrayed in Figure 1-1—the gap between aspirations and fulfillment—are also involved in revolutions. And they might give us further insight into relationships between aspirations and fulfillment, such as conditions under which that gap would increase or decrease. Further, we can also look to the impact that revolutions have on that gap. What can they tell us? Yet there are a great many other kinds of studies, a few of which will be included here, which can yield very useful information on such questions. Our web approach enables us to patch together bits and pieces of knowledge that ordinarily are kept separate from one another. And the result is the kind of credible knowledge that far exceeds the narrow base of knowledge on which specialized experts rely. That still will not enable us to predict revolutions, but perhaps it will help us to develop conditions that will aid in solving the problems that produce revolutions.

Revolutions

Merton's analysis of **anomie** emphasized the same forces portrayed in Figure 1-1, namely, the failure of "institutionalized means" to fulfill "cultural goals" or **cultural values.** He suggested five "modes of adaptation" to this situation, which we have analyzed as fundamental to the **social structure** of modern society: **"conformity,"** "innovation," "ritualism," "retreatism," and "rebellion." His category of "innovation," where institutionalized means fail to fulfill **cultural values** or goals, has been the most influential one among sociologists, having an important impact on studies of crime and **deviant behavior.** Yet his category of "rebellion" also appears to be quite useful in understanding the forces behind revolutions. Within this category there is both a rejection of traditional cultural goals and institutionalized means and also the development of alternative ones:

> When the institutional system is regarded as the barrier to the satisfaction of legitimized goals, the stage is set for rebellion as an adaptive response. To pass into organized political action, allegiance must not only be withdrawn from the prevailing **social structure** but must be transferred to new groups possessed of a new myth. The dual function of the myth is to locate the source of large-scale frustrations in the **social structure** and to portray an alternative [social] **structure** which would not, presumably, give rise to frustration of the deserving. It is a charter for action. (1949:145)

Merton's argument here is similar to what has been portrayed within Figure 1-1. Rebellions or revolutions, and not just crime, are based on the gap between aspirations and fulfillment. They are based on situations where the existing **social structure** comes to be seen "as the barrier to the satisfaction of legitimized goals." It is, then, the combination of high aspirations and their failure to be fulfilled that yields the gap that is the basis for rebellion or revolution. Further, revolutions also require "a new myth," just as Kuhn suggested that a new scientific paradigm is required if the old one is to be replaced. In the area of cultural and not just scientific revolutions, this suggests the necessity of a new **worldview** if the old one is to be replaced. What our own analysis adds to Merton's, as portrayed in Figure 1-1, is a view of industrialization and modernization as creating and escalating the gap that Merton is discussing. From this perspective, the very problems that are so fundamental as to be the basis for rebellions and revolutions are increasing as modernization proceeds. We can choose, following the Springdalers, to avert our eyes from those problems, making them relatively invisible at least until they explode. Or we can make those problems visible and consider how the social structure of modern society might be changed so as to resolve those problems. If we choose the latter course, following Merton, "allegiance must not only be withdrawn from the prevailing **social structure** but must be transferred to new groups possessed of a new myth" (1949:145).

Merton's analysis of rebellions as requiring a failure to satisfy "legitimized goals" is similar to the analysis of the French Revolution made by Alexis de Tocqueville in the middle of the nineteenth century:

> In 1780 there could no longer be any talk of France's being on the downgrade: on the contrary, it seemed that no limit could be set to her advance. And it was now that theories of the perfectibility of man and continuous progress came into fashion. Twenty years earlier there had been no hope for the future; in 1780 no anxiety was felt about it. Dazzled by the prospect of a felicity undreamed of hitherto and now within their grasp, people were blind to the very real improvement that had taken place and eager to precipitate events. (1955:177; quoted in Davies 1971:96)

Following de Tocqueville's argument, the French Revolution was largely based on very high expectations for improvement, corresponding to the top curve of rising expectations in Figure 1-1. As a result, "people were blind to the very real improvement that had taken place," implying that upward movement of the bottom curve made little difference if the very high expectations of the top curve remained unfulfilled. The Enlightenment had given the French people "the prospect of a felicity undreamed of hitherto," and they were unwilling to settle for less.

For further analysis of revolutions, let us turn to the twentieth-century historian Clarence Crane Brinton (1952), who drew this conclusion after analyzing the Puritan, American, French and Russian revolutions:

> First, these were all societies on the whole on the upgrade economically before the revolution came, and the revolutionary movements seem to originate in the discontents of not unprosperous people who feel restraint, cramp, annoyance, rather than downright crushing oppression. Certainly these revolutions are not started by down-and-outers, by starving, miserable people. . . . These revolutions are born of hope, and their philosophies are formally optimistic. Second, we find in our prerevolutionary society definite and indeed very bitter class antagonism. (p. 318)

Brinton agrees with de Tocqueville on the importance of positive changes, writing that those revolutions were "born of hope" and carried forward by "not unprosperous people," suggesting once again the top curve of rising expectations in Figure 1-1. He then adds to this idea feelings of "restraint, cramp, annoyance" coupled with "bitter class antagonism," suggesting the limitations placed on those hopes by the bottom curve of Figure 1-1 which depicts the fulfillment of expectations.

The political scientist James C. Davies attempted to combine various theories of revolution in his own research:

> Revolutions are most likely to occur when a prolonged period of objective economic and social development is followed by a short period of sharp re-

versal. People then subjectively fear that ground gained with great effort will be quite lost: their mood becomes revolutionary. The evidence from Dorr's Rebellion, the Russian Revolution, and the Egyptian Revolution supports this notion: tentatively, so do data on other civil disturbances. . . . The notion that revolutions need both a period of rising expectations and a succeeding period in which they are frustrated qualifies substantially the main Marxian notion that revolutions occur after progressive degradation and the de Tocqueville notion that they occur when conditions are improving. By putting de Tocqueville before Marx but without abandoning either theory, we are better able to plot the antecedents of at least the disturbances here described. (1962:5, 17)

Davies's approach is much like that of Brinton, combining the de Tocqueville notion of improvement, the top curve of Figure 1-1, with the Marxist idea of degradation. And if we come to see our "revolution of rising expectations" as linked to the very processes of industrialization and modernization, then the conditions that make for revolutions are continuing to increase while efforts to confront those conditions remain virtually nonexistent.

One promising approach to understanding the forces producing the growing gap between aspirations and fulfillment in modern society makes use of the situational concept of **relative deprivation**. In a study of 114 examples of political strife in areas throughout the world, Ted Gurr suggests the importance of this concept for understanding those scenes:

> The basic theoretical proposition is that . . . relative deprivation . . . is the basic precondition for civil strife of any kind, and that the more widespread and intense deprivation is among members of a population, the greater is the magnitude of strife in one or another form. Relative deprivation is defined as actors' perceptions of discrepancy between their value expectations (the goods and conditions of the life to which they believe they are justifiably entitled) and their value capabilities (the amounts of those goods and conditions that they think they are able to get and keep). . . . The underlying causal mechanism is derived from psychological theory and evidence to the effect that one innate response to perceived deprivation is discontent or anger, and that anger is a motivating state for which aggression is an inherently satisfying response. . . . The fundamental proposition that strife varies directly in magnitude with the intensity of relative deprivation is strongly supported. . . . Deprivation attributable to such conditions as discrimination, political separatism, economic dependence, and religious cleavages tends to contribute at a relatively moderate but constant rate to civil strife whatever may be done to encourage, deter, or divert it, short only of removing its underlying conditions. (1968:1105, 1123–24; see also 1970)

Gurr goes on to distinguish among three different routes to feelings of relative deprivation that foster revolution, all based on the relationships between curves of aspiration and curves of fulfillment or achievement. There is what he calls "aspirational deprivation," where achievement remains relatively constant but aspirations grow. As a result there is a revolution of rising expectations, and people "are angered because they feel they have no means for attaining new or intensified expectations" (1970:50). This is the situation depicted in Figure 1-1. Then there is the symmetrical opposite of aspirational deprivation, namely, "decremental deprivation," where the top curve of aspirations remains relatively constant or horizontal but there is a sudden drop in the bottom curve for the fulfillment or achievement of aspirations, such as in the case of a depression or economic crisis. This is what Gurr calls the "revolution of withdrawn benefits." Finally, there is the combination of these two phenomena, following Davies's emphasis, where there are both rising expectations and declining abilities to fulfill them, or what Gurr calls "progressive deprivation." In all three cases there is a growing gap between the curves of aspiration and their fulfillment. Never mind whether that gap is created by increasing aspirations, decreasing fulfillment, or some combination of the two: the result is increasing relative deprivation that points toward revolution.

The social science literature on revolutions is divided among those emphasizing social psychological explanations, like Davies and Gurr, and those emphasizing **social organization,** like Skocpol (1979) and Tilly (1978). For example, Skocpol looks to political and economic breakdowns within the old regime that create opportunities for revolt, and also to the development of new state organizations that can take charge while using revolutionary symbols. And Tilly centers on social organization within the political domain. He sees revolutions as an extreme form of the conflict for political control of the state, where the revolutionary group is able to command the resources required to take power from the old regime. Although the literature on revolutions is divided among those emphasizing social psychological explanations and those emphasizing different approaches to patterns of social organization, that division appears to be a function of a bureaucratic approach to the scientific method rather than the nature of the beast. A number of analysts have implicitly adopted aspects of our own web approach to the scientific method, where it is important to take into account **social structure,** the **individual,** *and* the situation (see, for example, Sayles 1984; Taylor 1984; Himmelstein and Kimmel 1981; and Sztompka 1993:301–21). From this perspective, there is a great deal that we can learn from all of these studies; they need not be seen as mutually contradictory.

Yet if these studies teach us to understand the nature of those forces generating basic and increasing problems in modern societies, what can we learn about the impact of revolutions on those problems so that we might gain understanding of how to address them more effectively? In his review of the literature on revolution, Sztompka concludes:

> Revolutions, especially when successful, engender heroic myths; their ac-
> complishments are exaggerated, the costs ignored. But from some historical
> perspective the side-effects, the human price, the boomerang effects, become
> unraveled, tempering the early euphoria. Quite soon the heroic myth of the
> Russian Revolution was crushed by the evidence of misery, oppression, sav-
> agery and death that it brought about. The final collapse of communism at
> the end of the twentieth century provided the ultimate proof that the project
> it attempted to implement was entirely misconceived from the outset. Then
> there is the heroic myth of the great French Revolution crumbling under the
> evidence provided by recent "revisionist" historiography (Sullivan 1989;
> Schama 1989), and recently ironically referred to as "so glorious, yet so sav-
> age" (Sullivan 1989:45). Why is it so often the case that revolutions produce
> something so utterly different from what was dreamed of by the revolution-
> aries? Why is it that the momentum of the revolution so often "demolishes
> so ruthlessly that in the end it may annihilate the ideals that called it into
> being" (Kapuscinski 1985:86)? Is this vicious logic inescapable? We do not
> know. (1993:319)

One response to Sztompka's queries about our lack of knowledge as to what causes the failures of revolutions is that in fact we *do* know something about those causes. For example, we might turn to Robert Michels's (1949) analysis of the socialist movement in Germany at the beginning of the twentieth century, a social movement and not a revolution but revealing nevertheless. He found that as the socialist movement gained adherents and became more formalized, with a bureaucratic pattern of organization replacing its initial loose and informal organization, something happened to its original reformist ideals. Those ideals gradually became displaced by the goals of maintaining the existence of the organization as well as strengthening it. As a result of his research, Michels formulated what has come to be known as "the iron law of oligarchy." This is the principle that every large organization, no matter how egalitarian its ideals, must estab-lish a **bureaucracy** with a few leaders monopolizing power if it is to put its program into effect. It is the presumed inevitability of this occurrence that makes it an "iron law." We might revise this so-called iron law and elimi-nate its inevitability by claiming that within a **bureaucratic worldview** where no other realistic options exist, the iron law summarizes a good deal of evidence relating to both social movements and revolutions. For exam-ple, Stalin's "temporary" dictatorship of the proletariat became perma-nent.

More generally, we can learn that if we adopt an interactive approach to the scientific method and move toward an **interactive worldview**—as was sketched in the above section—then there is a great deal to be learned about our experiences with revolutions that helps to explain their failures. Overall, as we have seen in Chapter 3, the major social problems of modern society—such as **anomie, alienation, social stratification** within the context of egalitarian **cultural values,** and **relative deprivation**—must be confronted for most revolutionary ideals to be fulfilled. However, those relatively invisible problems are all linked together and also tied to a wide range of relatively visible problems. Further, following our overall analysis in earlier chapters, all of these problems are tied to a number of other forces, as illustrated by the concepts within Figure 1-3, such as **worldview, bureaucracy, labeling,** and **addiction.** Even sociologists and other social scientists have yet to move beyond a bureaucratic approach to the scientific method so that they can make visible a wide range of these invisible forces and take them into account, given their own commitments to a **bureaucratic worldview** that holds a narrow approach to the scientific method in place. Following the above section, they continue to yield anomie, alienation and addiction. Under these conditions, it is quite understandable that revolutionaries with their bureaucratic worldview will be unable to understand the complex forces involved in achieving fundamental social change, let alone effecting such change.

Nevertheless—provided we adopt an interactive approach to the scientific method—we can learn a great deal from the literature on revolutions that can teach us how to confront basic social problems. For one thing, we sociologists can learn to take far more seriously situational forces like relative deprivation, given the substantial evidence for its importance. If the momentary scene is important in understanding the onset of revolutions, then it is probably also important for solving the problems that revolutions address. More specifically, we can come to understand the three routes to feelings of **relative deprivation** outlined by Gurr: aspirational deprivation, decremental deprivation, and progressive deprivation. Further, if social structure along with the structures of the individual should also be taken into account to understand revolutions, then we need no longer limit ourselves to studies of revolution in order to understand revolution. For we can learn about those structures—along with the situational forces illustrated by relative deprivation—within almost all sociological studies, and we can apply what we learn to an understanding of revolutions. Apparently, the study of revolutions need no longer remain like the study of earthquakes, where our failure to predict them is also the failure of sociology. The study of revolutions can teach us, for example, about how to fulfill far-reaching revolutionary ideals such as those which were sketched in Figure 1-5.

Satyagraha

One thing that many sociologists, anthropologists, and other social scientists have taught us over the years is the importance of **cultural values,** granting that their emphasis has been more on cultural differences than on cultural universals, such as a **bureaucratic worldview** based on patterns of **cultural conformity, social stratification,** and **bureaucracy.** Another example is the **cultural value** of "equality," which has come to be associated with the modern world. To the extent that such a universal worldview is found to exist, then it would give all peoples far more of a basis for learning to interact peaceably with one another by making visible the relatively invisible factors that they share. Donald Brown, a contemporary anthropologist, has commented forcefully on the self-serving forces that propel anthropologists to avoid emphasizing cultural universals:

> What anthropologists have to say about humanity has incalculable consequences for the peoples they study and for the public they report to. . . . The more those [sociocultural] differences can be shown to exist . . . the more sociocultural anthropologists (or sociologists) can justify their role in the world of intellect and practical human affairs and thus get their salaries paid, their lectures attended, their research funded, and their essays read. . . . Anthropologists . . . are the ones who reported stress-free adolescence among Samoans . . . and timelessness among Hopi—or who accepted these reports and wove them into a mythology. . . . This more than anything else lent the weight of empirical science to those extreme forms of relativism that hold or lead to the position that there are virtually no pancultural regularities or objective standards. (1991:154–55)

Brown has reference here to studies in anthropology challenging earlier work emphasizing the diversity of cultures at the expense of any attention to universals. For example, Margaret Mead's *Coming of Age in Samoa* (1928), based on nine months of fieldwork not preceded by learning Samoan, argued for a great difference between Western culture's approach to adolescence with its resultant stresses for young people and the Samoans' supposedly stress-free adolescence produced by a much different culture. However, Derek Freeman's *Margaret Mead and Samoa: The Making and Unmaking of a Myth* (1983; see also Freeman 1989), based on six years of fieldwork, found otherwise. In other work pointing away from cultural universals, Benjamin Lee Whorf and Edward Sapir formulated the "linguistic relativity hypothesis": people speaking different languages will come to understand the world in different ways (Sapir 1929; Whorf 1963). Fundamental to their conclusion was Whorf's argument that the Hopi Indians either had no sense of time or that they viewed time very differently from the way we do. But Malotki's (1983) thorough analysis of the Hopi documents the richness of their sense of time as well as the similarity of their conception to our own. The key implication of Freeman's and Malotki's work is not to suggest that cultural differences are unimportant but to suggest that cultural universals are fundamental.

This controversy within the literature of anthropology parallels the history of sociology. The anthropological emphasis on cultural differences justifies the status of anthropology as a distinct discipline, which, by comparing exotic cultures with our own, can educate us about the supposedly arbitrary yet very powerful impact of culture. It is a view of culture that enables anthropology to hold its own against the claims of biology and psychology about other sources of human behavior. Sociology's own status has been justified, in parallel, by reference to the overriding importance of patterns of social organization like stratification and bureaucracy as over against the individual. We might recall here the criticism of this one-sidedness voiced by Dennis Wrong in his "The Oversocialized Conception of Man in Modern Sociology" (1961) and by George Homans in his "Bringing Men Back In" (1964). All of this suggests the existence among anthropologists no less than sociologists of a **bureaucratic worldview** with its emphasis on **social stratification, bureaucracy,** and **conformity.** Yet it also suggests the possibility of a greater emphasis not only on cultural universals but also on the possibilities offered by an **interactive worldview,** which would also open up to a wide range of invisible forces that all humans share. And that in turn would open up to possibilities for utilizing what we share as a basis for effective policies in international relations.

One dramatic illustration of those possibilities was Mohandas K. Gandhi's successes in leading India toward independence from British rule. By centering on this one particular revolutionary struggle we can gain the concreteness that our general examination of revolutions in the above subsection lacks. Within a **bureaucratic worldview** with its seesaw metaphor, power is a zero-sum game, a fixed pie of rewards, where the gain by some is at the expense of others. Indeed, most of political sociology is oriented in this way, which surely would have received the blessings of Niccolò Machiavelli. Another alternative, following an **interactive worldview,** is to conceive of the development of power based on influence more than on force, where influence rests on legitimation through making visible shared **values,** which previously had remained largely invisible, and embodying those values. We might focus in particular on Gandhi's struggle in 1930–1931 in opposing the British Salt Acts. Those laws gave Britain a monopoly over salt production, preventing Indians from making their own salt. They worked financial hardships on the poor and symbolized unrepresentative government. This struggle or *satyagraha* by Indians became part of a year-long civil disobedience movement, with headquarters in Bombay, where activities were launched in every province (Bondurant 1965:88–102).

Earlier campaigns, despite difficulties encountered, had prepared the way for the successes of the salt *satyagraha*. For example, the Ahmedabad labor struggle in February and March 1918 was a dispute between textile workers and mill owners as to the amount of a cost-of-living increase to be paid to workers following the withdrawal of a high "plague bonus" after danger from plague had subsided. Initially, Gandhi was called in as one of three arbiters and determined that a 35 percent increase was justified by the higher cost of living. The mill owners, who previously had offered 20 percent, refused, and as leader of the workers Gandhi employed his techniques of *satyagraha*. Those procedures included the principle of self-sufficiency, where laborers would earn a living by undertaking other labor during the strike, even though such labor might appear to be demeaning or was a radically different experience for a worker. The *satyagrahi*s were trained to avoid violence against strikebreakers and anyone else, to hold fast to their resolve and not surrender no matter how long the strike lasted, and not to engage in "mischief, quarreling, robbing, plundering, or abusive language or cause damage to mill-owners' property, but to behave peacefully during the period of the lockout" (ibid.:68). Finally, the mill owners agreed to return to the arbitration table, and the dispute was ultimately settled by an increase of 35 percent. Emerging from the struggle was the Ahmedabad Textile Labor Association, the most powerful labour union in the country with a membership of 55,000, devoted to nonviolence principles as well as to constructive welfare work.

Reacting to the British Salt Acts, Gandhi together with other leaders planned a two-hundred-mile march to the sea, from Ahmedabad to Dandi, where volunteers would proceed to prepare salt from sea water. Volunteers accepted this pledge:

1. I desire to join the civil resistance campaign for the Independence of India undertaken by the National Congress.
2. I accept the Creed of the National Congress, that is, the attainment of *Purna Swaraj* [complete independence] by the people of India by all peaceful and legitimate means.
3. I am ready and willing to go to jail and undergo all other sufferings and penalties that may be inflicted on me in this campaign.
4. In case I am sent to jail, I shall not seek any monetary help for my family from the Congress funds.
5. I shall implicitly obey the orders of those who are in charge of the campaign. (ibid.:92)

Prior to the launching of the campaign, the movement for complete independence was advanced through widespread discussion throughout India as well as the deliberations of the Congress party. Training courses were initiated for volunteers who would participate in the salt *satyagraha*.

Vallabhbhai Patel was chosen by the National Congress, then headed by Jawaharlal Nehru, to prepare the people along the route of the proposed march. He advised them of the objectives of the campaign, teaching them the principles of *satyagraha*. He pleaded with them to undertake constructive work, not to drink intoxicants, and to overcome their patterns of discrimination against Untouchables. On March 7, 1930, he was arrested by the police. In a letter delivered March 2 to Lord Irwin, the British viceroy, Gandhi reviewed the specific grievances of the people of India and the aims of the salt *satyagraha*. He also told of the specific plans for the march, including March 12 as the date of its initiation, and he urged further discussion and a negotiated settlement. He stated:

> It is, I know, open to you to frustrate my design by arresting me. I hope that there will be tens of thousands ready, in a disciplined manner, to take up the work after me, and, in the act of disobeying the Salt Act, to lay themselves open to the penalties of a Law that should never have disfigured the Statute Book. (ibid.:93)

Gandhi was not arrested at that time, and on March 12 he and his co-*satyagrahi*s left Ahmedabad, attracting nationwide attention and reaching Dandi on the coast on April 5.

After prayers the following morning, Gandhi and his followers proceeded to the beach at Dandi, where they prepared salt from the sea water and, as a result, broke the salt laws. This was followed by acts of civil disobedience throughout India. Gandhi declared to the press that anyone willing to risk prosecution should make salt in defiance of British law. The Congress party published leaflets containing instructions on how to manufacture salt and distributed them throughout the country. Nehru wrote that people everywhere were beginning to make salt, collecting salt water in pots and pans, and ultimately succeeding in producing "some unwholesome stuff":

> As we saw the abounding enthusiasm of the people and the way salt-making was spreading like a prairie fire, we felt a little abashed and ashamed for having questioned the efficacy of this method when it was first proposed by Gandhiji. And we marveled at the amazing knack of the man to impress the multitude and make it act in an organized way. (ibid.:94)

Throughout the country shops closed in response to the arrest of *satyagraha* leaders, and headmen in villages along with subordinate officers resigned in large numbers. Nonpayment of taxes was undertaken in some areas. New leaders took the places of those had been arrested. Nehru, arrested on April 14, was succeeded by his father.

One problem Gandhi faced was how to preserve the nonviolent character of his campaigns. For example, during the Ahmedabad campaign he had to resort to a fast when the strikers started menacing the strike-breakers and the campaign hinted at weakness. He declared: "Unless the strikers rally and continue the strike till a settlement is reached, or till they leave the mills altogether, I will not touch any food" (ibid.:68). During the extension of acts of civil disobedience throughout the country following the march to the sea, there were outbreaks of riots in Karachi and Calcutta. Gandhi declared on April 17, "If non-violence has to fight the people's violence in addition to the violence of the Government it must still perform its arduous task at any cost" (ibid.:95). During the first week of May he sent a second letter to the British viceroy explaining his next move: to set out for Dharsana and demand possession of the large salt works located there. He stated that this "raid" could be prevented by removing the salt tax or by arresting or assaulting Gandhi and all of his followers. Gandhi was arrested on May 5, but leading Congress officials at the head of volunteers marched to Dharsana to occupy the salt depots. As many were struck down, fresh volunteers stepped in to replace them, and first-aid units organized by the leadership worked to revive the victims.

All throughout the raid, volunteers did not strike back at those who struck them down, and they even did not deflect those blows. Wave upon wave, they rushed to occupy the salt pans, sometimes pleading with the police to join them. And there were incidents of police who refused to continue their assault on the volunteers:

> An American journalist, Negley Farson, recorded an incident in which a Sikh, blood-soaked from the assault of a police sergeant, fell under a heavy blow. Congress first-aid volunteers rushed up to rub his face with ice. . . . " He gave us a bloody grin and stood up to receive some more." . . . The police sergeant was "so sweaty from his exertions that his Sam Browne had stained his white tunic. I watched him with my heart in my mouth. He drew back his arm for a final swing—and then he dropped his hands down by his side. 'It's no use,' he said, turning to me with half an apologetic grin. 'You can't hit a bugger when he stands up to you like that!' He gave the Sikh a mock salute and walked off." (ibid.:96)

After the monsoon season started, the salt *satyagraha* was replaced by other campaigns, such as an economic boycott of foreign-made products and civil disobedience of special ordinances that were designed to suppress and control the assembly of participants in the movement. The campaigns continued throughout 1930, involving extensive noncooperation and civil disobedience.

Finally, after talks between Gandhi and Lord Irwin, the British viceroy, the Gandhi-Irwin Agreement was published on March 5, 1931. Although the salt laws were not repealed, a new interpretation made it possible for poor people to obtain relief from the salt tax. Specifically, local residents in villages adjacent to areas where salt could be made would be allowed to make salt for consumption within those villages. Lord Irwin also granted amnesty to all individuals convicted of nonviolent offenses related to civil disobedience and the restoration of all confiscated or forfeited properties. Further, representatives of the Congress party would be invited to partic-ipate in the deliberations of the next Round Table Conference on such ques-tions as federation, financial credit, defense, and the position of minorities. Gandhi, on his part, agreed to end civil disobedience, such as organized defiance of the law like the salt campaign, nonpayment of legal dues, pub-lication of pamphlets supporting civil disobedience, and efforts to influ-ence officials to work against the government or to resign. Gandhi and the Congress party did not press for independence during World War II, given the Nazi and Japanese threats, and its loyalty was finally rewarded just af-ter the war. Britain lost control of India's foreign policy, but India retained its cultural ties with Britain, continued to trade with her, supported dem-ocratic forms of government, no longer required Britain to maintain a large military force in India or to experience a devastating **value** conflict.

In looking to Gandhi's achievements in the first half of the twentieth cen-tury from the perspective of the twenty-first century, and in particular his approach to *satyagraha*, our focus remains the same as at the beginning of this chapter. How does *satyagraha* help us to understand the nature of an interactive sociological paradigm and **worldview**? What insights might be involved? And what can it say to us at this time in history about the fun-damental and escalating problems we are facing? George Lundberg de-scribed our modern situation in 1961:

> A leader, however admirable in ability and intentions, attempting to admin-ister centrally a large society today is somewhat in the position of a pilot try-ing to fly the modern stratoliner without an instrument board or charts. . . . Only as a result of the development of the basic physical sciences can a large modern airplane either be built or flown. Only through a comparable devel-opment of the social sciences can a workable world order be either con-structed or administered. The appalling thing is the flimsy and inadequate information on the basis of which even a conscientious executive of a large state is today obliged to act. (ibid.:142)

As we attempt to fly our ship of state we are confronting increasing prob-lems, based on both the snowballing of the physical and biological sciences and their technologies as well as invisible forces such as anomie and alienation linked to the failures of the social sciences. These were illustrated schematically in Figures 1-1 and 1-4. How can we learn to address those problems?

Satyagraha has been interpreted within the literature of conflict resolution as involving at least these eight elements:

1. *Refraining* from any form of verbal or overt *violence* toward members of the rival group.

2. Openly *admitting* to the rival group one's plans and intentions.

3. *Refraining* from any action that will have the effect of *humiliating* the rival group.

4. Making visible *sacrifices* for one's cause.

5. Maintaining a consistent and persistent set of *positive activities* which are explicit (though partial) realizations of the group's objectives.

6. Attempting to initiate direct personal *interaction* with members of the rival group, oriented toward engaging in *friendly verbal discussions* with them concerning the fundamental issues involved in the social struggle.

7. Adopting a consistent attitude of *trust* toward the rival group and taking overt actions which demonstrate that one is, in fact, willing to act upon this attitude.

8. Attempting to achieve a high degree of *empathy* with respect to the motives, affects, expectations, and attitudes of members of the rival group. (Janis and Katz 1959:86)

The foregoing description of the salt *satyagraha* along with the brief reference to the Ahmedabad campaign illustrate most of these elements. For example, Gandhi's letters to Lord Irwin suggested discussions between the two (element 6) and indicated his specific plans and intentions, including the dates when he would march to the sea and when he would march on the Dharsana salt works (element 2). As for positive activities (element 5), the principle of self-sufficiency required strikers at Ahmedabad to find other work, and during the march to the sea within the salt *satyagraha* people along the route were urged to "undertake constructive work" and to "overcome their patterns of discrimination against Untouchables." As for element 1, refraining from violence was central to the whole approach, which required that all volunteers adhere to the Creed of the National Congress, that is, the attainment of complete independence "by all peaceful and legitimate means." When violence was threatened at Ahmedabad, Gandhi initiated his fast. When there were outbreaks of riots in Karachi and Calcutta, Gandhi declared: "If non-violence has to fight the people's violence in addition to the violence of the Government it must still perform its arduous task at any cost." (quoted in Bondurant 1965: 95). And the march on Dharsana, despite the enormous cost in human life and injury, was nonviolent. It also illustrates extreme personal sacrifices (element 4)

As for attempting to achieve empathy with the rival group (element 8), Gandhi insisted to the mill-owners at Ahmedabad that they should not let his fast influence them—since he knew them personally—but that they should be free to make their own decision, and that it was not part of an ideal *satyagraha*. We can see it as violating element 3, where the rival group is humiliated, granting that the overall struggles at Ahmedabad and within the salt *satyagraha* avoided such humiliation. We might also note empathy (element 8) within the texts of Gandhi's letters to Lord Irwin, where he takes Irwin's position and spells out Irwin's choices and their possible consequences. As for trust (element 7), perhaps the best example of this was not illustrated above. It was Gandhi's willingness to defer the struggle for independence, which was fundamental to all of the *satyagraha* campaigns, until after World War II. But such trust is also illustrated by many of the specific elements listed above and not just in element 7. For example, letting your rival know your specific plans and intentions (element 2) requires at least some faith in your rival. And attempting to achieve empathy with your rival (element 8) suggests that you are treating him or her as a human being much like yourself, and if you can trust yourself you might also learn to trust your rival.

If we look to our twenty-six sociological concepts, we can find Gandhi's procedures for *satyagraha* as illustrating an **interactive worldview** and opposing a **bureaucratic worldview**. For example, egalitarian **social interaction** is illustrated by attempting to initiate personal interaction with the rival group(element 6), and also by admitting one's plans and intentions (element 2). Also, avoiding negative **reinforcement** as well as avoiding **social stratification** are exemplified by refraining from violence (element 1) and refraining from humiliating the rival group (element 3). The positive activities emphasized in element 5 illustrate positive **reinforcement**. And these efforts to avoid negative reinforcement and social stratification while using positive reinforcement would work toward reducing feelings of **relative deprivation** within the rival group. At the same time, personal interaction (element 6), revealing plans and intentions (element 2), trust (element 7), and empathy (element 8) suggest an emphasis on **cultural values** and **cultural norms** that support an interactive worldview, such as "equality," "freedom," "individual personality," and "democracy." We might also invoke **social organization** in the commitment of the Congress party to principles of nonviolence and to teaching the **groups** of volunteers **conformity** to those principles so that they no longer continued **conformity** to patterns of **social stratification**. Those **groups** included representatives from the entire spectrum of the Indian population, such as untouchable Hindus, Brahmans, Muslims, Pathan warriors, laborers, peasants, the educated, and the wealthy.

Fundamental to the nature of *satyagraha* is the effort to employ means that do not sacrifice the end in view, by contrast with what we have experienced in Marxism, as stated by Bondurant:

> In the realm of political philosophy, as on the field of action, the dynamic technique of satyagraha suggests a re-examination of the means-ends relationship. *Satyagraha*, claiming to be more than means, to be, indeed, end-creating, introduces a dynamic element with challenging implications for political method. If the dichotomy of ends and means is yet meaningful, *satyagraha* confronts the cry of Spengler that man needs above all a noble end, with the inherent proposition that what political man needs is not a noble end, but constructive, creative means. (Bondurant 1965:231)

We might see ends as encompassed by **cultural values** as well as **worldview** and means as illustrated by the range of situational ideas, feelings, and actions, as illustrated by **definition of the situation, label, relative deprivation, reinforcement, conformity, deviance,** and **social interaction.** Although Bondurant states that "what political man needs is not a noble end, but constructive, creative means," an **interactive worldview** suggests that political man needs the two interacting with one another, that is, the interaction between structures and situations.

More concretely, it was Mohandas Gandhi who was able to give life to this abstract idea, risking his life at every crossroads to achieve independence for India, and it was also volunteers like the Sikh who stood up for more punishment after his head had been battered and bloodied by a policeman's nightstick. We do not need to subscribe to a great-man theory of history to pay our dues to the importance, for the successes of the *satyagraha*, of the **individual, personality structure, self-image,** and **worldview.** To pay attention to these forces is of course not to understand much about them. But it is at least to give recognition to a wider range of forces involved in the Indian revolution than a focus limited to **social structure** and the situation would permit. Even given all of the analyses of Gandhi's life, we still know very little about exactly what combination of experiences and forces resulted in this individual who was able achieve so much. Yet there is much to learn about the nature and impact of those forces, on other individuals in other situations and at other times, from the literature of sociology. Our focus continues to be on interaction as one key to understanding the nature of those forces. Another key is the emotional life of the **individual,** but taking into account as well social structure and the situation.

The Individual and Awareness of Emotions

We referred in Chapter 2 to Thomas Scheff's (1997) part-whole analysis as paralleling the approach to the scientific method adopted in this book.

Our focus here will be on his specific analysis of shame and other emotions within the context of efforts at family mediation. Although this topic appears to be a rather specialized one, in fact it invokes our own overall approach as well as our particular direction in this subsection: to illustrate research within sociology that emphasizes interaction and succeeds in addressing the fundamental problems of modern society. There is a great deal of overlap between Scheff's orientation and our own, just as his part-whole analysis overlaps with what has been presented in earlier chapters. Our own interpretation of Scheff's work and that of several others will focus on employing the concepts presented in Chapter 1 as well as the figures presented in Chapters 1 and 4. In other words, it will emphasize our invisible crisis and how it might be confronted. The acknowledgment of shame is analogous to the acknowledgment of the existence of the escalating gap between aspirations and their fulfillment presented in Figure 1-1. Both individual shame and the existence of that gap appear to be largely invisible, and their hidden nature works to prevent us from confronting those problems, yet it is indeed possible to make those phenomena more visible.

The key illustration Scheff uses is based on the work of Retzinger (1991), a psychotherapist who has been deeply involved in mediation within family settings and who has a background in sociology. Their focus is not on ordinary conflict but on "protracted" or "interminable" conflict, namely, disputes that resist resolution. Such conflict suggests the existence of structures—both social and personality—which are involved in the conflict, just as Figure 1-1 is best understood initially in relation to conflicts within social structures. What Scheff achieves is the linking of situational description with structures operating within that situation, thus invoking implicitly the full range of concepts within Figure 1-3. A couple in Retzinger's study, Rosie and James, volunteered to have one of their arguments videotaped. At the time of the study Rosie and James were husband and wife, white, middle-class, married for eleven years, and thirty-two and thirty-five years old, respectively. James had completed four years of college and was employed in the forestry service and Rosie was an undergraduate. They were in a trial separation and had been discussing divorce. The argument was over James's ownership of a private airplane. In the following transcription, capitalized words indicate loud speech, horizontal lines are used for interruptions, and the number of seconds is indicated:

1. 23.25 **R:**   so what aspects of the plane do you want to talk
2. about?
3. **J:**   just airplanes in general it doesn't have to (impasse)
4. be
5. **R:**   oh
6. **J:**   specifically the one we have now (laugh)

7. 30.09 **R:**   no I wanna NARROW it RIGHT down TO that one
8. **J:**   because I don't plan on it being the last the
9. end of the line
10. **R:**   NO well I don't either not for you (quarrel)
11. **J:**   oh good
12. **R:**   no I wouldn't take your toy from you
13. **J:**   all right
14. **R:**   I sacrificed a LOT for you to have toys
15. (both laugh tightly)

---

16. 49.00 **R:**   but you didn't ask for it and I resent (impasse)
17. later and we're still going over it ok
18. **J:** Yea it (3) it uh goes back to another era. (1991:18–19)

Following Scheff's interpretation, it is useful to divide this transcript into the three sections separated by the horizontal lines: impasse, quarrel and impasse, where the impasses indicate the couple's standard way of avoiding direct confrontation and, as a result, avoiding the possibility of solving their problems. In the first impasse, lines 1–6, they both begin in a rational mode, complying with the researcher's request to discuss a topic that they argue about frequently. In lines 1 and 2 Rosie "opens the argument by asking a question in an ingratiating, childlike manner," although "both Rosie and James know that it is Rosie who has a problem with the plane" (ibid.:20). She takes this tack of avoidance instead of stating her feelings forcefully, such as "I hate the plane and wish we would get rid of it" (ibid.). Apparently James sees this avoidance and comes to believe that he can do likewise by talking simply about planes in general. He is interrupted in midstream by Rosie with her surprised "oh" on line 5 (a horizontal line was not used for that interruption), but he continues gamely to complete his sentence in line 6, ending with a laugh, perhaps anticipating the tense situation he and Rosie have produced and attempting to dispel that tension. Scheff interprets lines 1–6 as illustrating hidden shame: "Both show false smiles, and both are soft-spoken and evasive about the topic to be discussed. These are cues to shame" (ibid.:24).

For Scheff, this impasse illustrates one kind of **alienation,** namely, the "engulfment" of the individual's **personality** structure and the suppression of **individual** emotions, much like Durkheim's argument for altruistic or fatalistic suicide like hara-kiri in Japan:

In using unity language, both Rosie and James become me's (objects) to each other rather than I's (subjects). Instead of giving voice to their own desires and feelings, each suppresses them in a show of unity. . . . Engulfment leads to massive **conformity** since self is subjugated to the other person or to the group. Rosie's submissive, childlike passivity in her opening line is an example. A wife inviting a husband to represent her position, rather than speaking for herself, suggests that stereotyped male and female roles in our society are a manifestation of engulfment. In the stereotyped female role, a woman is socialized so that first as a daughter, and then as a wife, she is submissive, an empty vessel, to complement the stereotyped male role of father or husband, the dominating leader in the family. . . . Engulfment is a more subtle form of alienation [than isolation] because disconnection between persons arises from disconnection from self. . . . In responding to Rosie's opening question with a line about planes in general, James also abandons parts of himself: his desire to own a plane, the resentment he seems to feel about Rosie's opposition, and perhaps some guilt about his own opposition to her. Like Rosie, he hides his desires and feelings to show unity with her. (ibid.:29)

Scheff sees a second kind of **alienation** that he calls "isolation," illustrated by the quarrel phase of the argument in lines 7–15, the kind of alienation analogous to egoistic or anomic suicide for Durkheim. Instead of the silent impasse that characterized the couple's normal mode of conflict, as in lines 1–6, we have the active quarrel. When James fails to focus on their own plane in lines 3–4 her surprised "oh" anticipates the anger she expresses in lines 7, 10, 12, and 14. We have nonverbal cues for this anger as well, since Retzinger videotaped the quarrel:

There are many indications of anger in Rosie's response to James's statement. Her interruption in line 5, her very quick response to line 6, and the rapidity of her speech are one set of indications. The flat contradiction in her "No," her wish for the opposite of his suggestion, the loud intonation given to some of the words, an aggressive leaning forward as in a challenge, the narrowing of her eyes, and the lowering of her eyebrows (as reported by Retzinger) all suggest anger. The beginning of line 10 starts with agreement . . . but by the end of the line, Rosie gives it a nonverbal twist that suggests separation from James. She drags out the last word "you" into two syllables, "you-ou," with the last syllable being the longest, her face expressing contempt and, more faintly, disgust. . . . In line 12 . . . she calls his airplane a toy. Beginning at this point, she is no longer offering him respect for an equal. . . . She treats him as a childish adult. (ibid.:21)

As for James, his response to Rosie's angry confrontation in line 7 is anger. The implication of lines 8 and 9 is that whatever Rosie wants no longer matters to him. But his overtly angry reaction changes in lines 11 and 13:

Before Rosie's first angry line (7), James had spoken in full sentence form. After line 7 the length of his responses decreases considerably: His two subsequent responses are all quite brief; each is only two words long. These two words suggest agreement with Rosie, but his manner contradicts his words.

In all three responses, he seems to be withdrawing from the conversation. His line 11 is delivered much more quietly than his earlier responses. Lines 13 and 18 continue to be soft. By line 13, his fixed smile has begun to fade, and there are subtle signs that he is shrinking down into his seat. (ibid.:21–22)

Whereas James shows anger in lines 8 and 9, he shows withdrawal in lines 11 and 13, given his two-word sentences and quiet speech. The term "passive aggression" has been applied to this kind of withdrawal. In effect, he is saying nonverbally, "I no longer need you. You are not even worth my getting angry with you." The difference between Rosie's and James's anger in the quarrel phase reflects a difference between the **socialization** of men and women, where men are taught to repress emotions more than women.

The "isolation" or conflict in lines 7–15 is followed by a second impasse in lines 16–18, which we can see foreshadowed by the tight laughter in line 15:

After line 14, the mood of Rosie's responses abruptly changes. Line 16 contains one word indicative of anger ("resent"), but it is now encased between conciliatory words and ideas: James did not ask for the sacrifice, and they are discussing it. Moreover, the manner of Rosie's delivery also changes from the sharp, loud staccato of lines 7–14 to an oversoft musing, almost as if addressed to herself rather than to James. . . . After line 14, it appears that Rosie has noticed the indications of James's withdrawal. As she explained to the interviewer, there is a brief bristle of anger; then one or the other backs down. Perhaps Rosie felt that her angry outbursts had hurt James's feelings or that he might break off the discussion if she persisted. At any rate, it is clear that by line 16, Rosie has swallowed her anger, changing to a much more distanced kind of rational discussion. After a brief confrontation, she has backed down. The couple has returned to the status quo, an impasse in which the issues that separate them are avoided (ibid.)

Overall, then, we have a sequence of impasse, quarrel, and impasse within this protracted or continuing conflict, indicating alternation between two kinds of **alienation** by contrast with solidarity: engulfment, isolation or conflict, and engulfment.

Scheff is able to draw a further contrast between the situations of engulfment and that of isolation by means of a close analysis of the specific words used within the two, a contrast that has implications for language usage within any situation whatsoever. For example, let us note Rosie's response (lines 16–17), which transforms the quarrel to an impasse: "But you didn't ask for it and I resent later and we're still going over it ok." The "you" comes before the "I" here, by contrast with her statements in the quarrel phase, making James the subject rather than herself, and the "and

I resent later" is mumbled and ungrammatical. Also, the last clause, "and we're still going over it ok" includes a "we" and the idea "ok," which appears to be a call for agreement. James's response begins with "Yeah," indicating the agreement that Rosie had asked for. This "unity language" was also used to some degree at the beginning of the first situation of engulfment, lines 1–3. Rosie's opening sentence in line 1, "so what aspects of the plane do you want to talk about?" has "you" as the subject with no "I" at all, and James's response centers on "airplanes" and "it" rather than "I." This contrasts markedly with the quarrel phase, where in lines 7, 10, 12, and 14 Rosie begins with an "I" as subject followed by a "you" as object and emphasizes the great difference between the two.

A key emphasis in Scheff's study is the hidden nature of emotions as well as the potential impact of making those invisible emotions visible, as revealed both in a debriefing session and a follow-up study:

> When the tape was paused the first time at the peak of Rosie's anger in the passage described here, Rosie pointed to her own image on the screen, saying, "*That* is one angry woman." She explained to Retzinger that at the time she was unaware that she was angry. She reacted in a similar way to other passages, as did James. When the tape was stopped toward the end of the passage described here, James turned to Rosie, saying, "*That's* the expression on my face that you have been telling me about." He was referring to his tense, withdrawing, shrinking look, which he called his "hurt" look. Apparently they were unaware of their own emotional expressions at the time of the quarrel. . . . Rosie and James extended the session (three hours rather than the usual one hour) and benefited from it. In a follow-up three years later, they were living together again. Rosie told the interviewer that the debriefing session had changed their lives. She also said that, although James still had his plane, it no longer stood between them. Participation in the study had led them in directions that had changed the relationship for the better. (ibid.:24–25)

What Scheff and Retzinger achieve in their work is both a direction for understanding some of the forces that prevent individuals from achieving a close relationship and also for understanding how those forces might be removed. And they accomplish this through a close analysis of both verbal and nonverbal patterns of interaction coupled with an abstract theoretical approach emphasizing the complexity of human relationships. More specifically, their contrast between "unity language" and the language of disagreement not only is highly specific but also can be used for analyzing language in general. A key aspect of their work was the uncovering of the generally hidden emotions of shame and anger, aided by videotaping. And basic to their success with Rosie and James was a long debriefing session where they were able to use that session to make visible to themselves the emotions that otherwise would have remained invisible. Here again, we can generalize these results to other situations, since this opportunity to view one's nonverbal behavior in this way constitutes a rare situation. The general implication here is that if the individual adopts a reflexive orientation, attempting to increase awareness of his or her own behavior, that this will help in the avoidance of isolation or conflict with others.

Scheff's theoretical and research background can help us to understand his approach more fully and also gain further insights. A key emphasis here is on the importance of shame and our failure to acknowledge it in modern society, as he illustrates here:

> The Old Testament contains many, many references to pride and shame but very few to guilt. The New Testament reverses the balance: There are many more references to guilt than to shame. One possible interpretation for this reversal is the difference between "shame cultures" and "guilt cultures" . . . (Benedict 1946). . . . This book, however, offers a different explanation. I argue that the distinction between shame and guilt cultures is misleading since it assumes that shame states are infrequent in adults in modern societies. It is possible that the role of shame in social control has not decreased but has gone underground instead. . . . For example, we say, "It was an awkward moment for me." This statement usually refers to a feeling of embarrassment. It contains two movements that disguise emotion: *denial* of inner feeling and projection of it onto the outer world. *I* was not embarrassed; it was the *moment* that was awkward. (Scheff 1994)

James gives us another example of the denial of shame in the debriefing session, when he referred to his tense, withdrawing, shrinking look during the quarrel phase of the argument as his "hurt" look: he was, in fact, ashamed.

Scheff takes us on a journey through several literatures dealing with shame, such as cross-cultural etymological analyses of the concept and its usages, biblical sources, Greek ideas and ideals, Darwin's analysis, the work of Cooley (1922), Goffman (1959), and Elias (1978), and his own earlier study (Scheff 1990), all in his chapter on pride and shame (Scheff 1994:39–55). He then focuses attention on research by the psychotherapist Helen Lewis (1971) with its emphasis on unacknowledged shame, a work that forms much of the basis of Scheff's analysis. Lewis analyzed the transcripts of hundreds of psychotherapy sessions moment by moment, finding that shame was by far the most prevalent of emotions expressed, outranking by far anger, grief, and fear. Further, almost all of these episodes of shame were unacknowledged by the patient as well as the therapist. Lewis went on to distinguish between "overt shame" and "bypassed" or "chronic shame." The former is illustrated by such disguises or code words as "awkward," "uncomfortable," "insecure," "stupid," or "rejected." Whereas patients feel unwanted physical symptoms such as blushing, sweating, or pounding heartbeat in overt shame, they feel little pain in bypassed shame, speeding up their thoughts and speech and distracting themselves by this rapid activity, which might be viewed as obsessive and compulsive.

Of special interest in Lewis's work is her analysis of the relations among shame, anger, guilt, and resentment or hate, where the former two (shame and anger) may be viewed as more situational than the latter two (guilt and resentment or hate), which may be seen as chronic or structural. She sees guilt as the product of a shame-anger sequence or loop, where shame is followed by anger, and anger is then followed by shame, and so on, just as in the sequence of the argument between Rosie and James, where the initial impasse was followed by their quarrel and then by a second impasse. She argues that these loops may be seen as self-perpetuating chains of emotions or "feeling traps," and that emotions can also loop back on themselves or one another indefinitely, such as when a patient is ashamed of being ashamed, and then is angry at being ashamed of that shame. And just as chronic guilt can result from such anger directed at oneself on repeated occasions, so can the patient come to feel chronic resentment or hate when that anger is directed outward. Scheff uses this argument in later chapters of his book on the origins of World War I and World War II in examining the long-standing resentment between France and Germany not only during the first half of the twentieth century but also during the latter part of the twentieth century (1994:75–126).

In a recent paper, Retzinger and Scheff (2000) explore the implications of their earlier work for mediation procedures, with a focus on the process of acknowledging hidden shame. To illustrate, Saposnek (1983) reported a custody dispute between "Joan" and "Paul," where "in the middle of a heated exchange, a wife said to her ex-husband, 'You never paid any attention to the children, then you left me, and you're not getting the children now or ever'" (p. 185). At that point the mediator intervened:

> The anger and hurt you feel right now is not unusual, and it is very understandable. It is also not unusual for a parent who was not involved with the children before a divorce to decide to become sincerely involved after the divorce. Allowing that opportunity will give your children a chance to get to know their father in the future in a way that you wanted in the past. But give yourself plenty of time to get through these difficult feelings. . . . The husband . . . then tearfully expressed his sincerity in wanting to become more involved with the children. The wife cried and was able to constructively express her hurt feelings at being left by the husband. (ibid.:185–86)

Retzinger and Scheff argue for the importance of the mediator's intervention: by interrupting the quarrel cycle and expressing shame and hurt for the clients, further escalation that would have made it difficult for the couple to bond as coparents was avoided.

We might note from the length of the foregoing exposition, based only on mere seconds of interchanges between couples, that a great deal is going on in human interaction, and that it is indeed difficult to understand all of this complexity. Yet Scheff and Retzinger manage not only to uncover crucial forces and problems involved but also to point up directions for solving those problems. By using videotapes and transcriptions within the context of theories from sociology and psychology, they are able to give us ways of understanding the significance of momentary occurrences—verbal and nonverbal—within human interaction, and not merely the specific sessions analyzed. Further, they are able to suggest links—through loops or "feeling traps"—between situational feelings like shame and anger and structural or chronic feelings like guilt and resentment or hate. As for addressing these problems, they suggest how mediators might be able to express hidden shame or other emotions for their clients and as a result make bonding more possible. Perhaps most significant, they point a direction for all of us to learn to use ordinary language in ways that will help us in our interactions with others. For example, Rosie and James can learn to recognize their overt and covert shame, just as Joan and Paul and the rest of us can.

Scheff and Retzinger, with their situational orientation and procedures, help us to see the importance of **relative deprivation** in understanding revolutions as well as Gandhi's *satyagraha* as a procedure that enables one's situational means to interact with one's structural ends. In both examples what we were attempting was an understanding of the enormous complexity of human behavior through an emphasis on the importance of situational concepts from the twenty-six portrayed in Figure 1-3. In order to unearth some of that complexity and use that understanding as a basis for solving problems, we can learn to combine the situational concepts—**definition of the situation, label, relative deprivation, reinforcement, conformity, deviance,** and **social interaction**—with structural concepts. Of course, structural concepts are important as well, and Scheff and Retzinger in their treatment of **alienation** do not neglect them.

However, we can build on their detailed work by bringing to bear on it several additional concepts having to do with both **social structure** and the **individual**. These concepts might help both mediators and all of us to unearth emotions from our interactions more easily and, as a result, deal with them more effectively so as to move toward the ability to create social bonds or, to introduce a twenty-seventh concept to our list, **social relationships** that are egalitarian or close. We might define **social relationship** simply as "a continuing pattern of social interaction." Scheff and Retzinger's distinction between "overt shame" and "bypassed shame," based on Lewis's work, is a basis for Scheff's distinction between the isolation or conflict in the quarrel phase of Rosie and James's argument and the en-gulfment in the impasse phases. Here it appears useful to introduce a twenty-eighth concept for dealing with the situational domination occur-ring in that quarrel phase as well as situational hierarchy—as distinct from structural **social stratification**—in general. We define **domination** as "the exercise of power or control over the behavior of others—against their will if necessary—in a given situation." For example, we might see the anger or rage expressed by Rosie and James in their quarrel phase as attempts at achieving **domination.**

Yet we can also use their analysis to bring in several other concepts that can help us to unearth even more of the complexities involved. For exam-ple, we can see Rosie's statement on line 14, "I sacrificed a LOT for you to have toys" as an instance of feelings of **relative deprivation.** And in that same sentence, her reference to James's "toys," which also occurs on line 12, is an indirect way of **labeling** James as an immature individual. We might also understand more fully James's reaction of withdrawal on lines 11 and 13 as the result of **negative reinforcement** coming from Rosie. What she is doing with her overt anger and what he is doing with his withdrawal and passive anger is failing to achieve egalitarian **social interaction.** All of this anger illustrates **deviance** from **cultural values,** supported by **cultural norms,** emphasizing "equality," "freedom," "individual personality" or the worth of every individual, and "democracy." Instead, they illustrate **conformity** to **cultural values** and **cultural norms** associated with patterns of **social stratification** and **bureaucracy.**

Returning specifically to the distinction between bypassed and overt shame, we can see it as illustrating the distinction between **addiction** and **social stratification.** On the one hand, all types of shame can be seen as a species of social stratification, a concept that can be applied to a variety of emotions, since a "persisting hierarchy" within a group can be evidenced by such varied phenomena as guilt, hate, pessimism, arrogance, haughti-ness, and selfishness. Addiction is defined as "the individual's subordina-tion of individuality to dependence on external phenomena." Along with bypassed shame, addiction is chronic, involves a good deal of **conformity** or subordination of the **personality structure,** and is characterized by obsessiveness and compulsiveness. Scheff and Retzinger's concern with **alienation** can be linked specifically to historical changes, such as those im-plied in Figure 1-1, with alienation increasing as modernization and **anomie** proceed. Their analysis, following that of Elias, of the continuing importance of shame and the idea that it has now gone underground, meshes with what sociologists have learned about the revolution of rising expectations. Along with industrialization and modernization we have in-creasingly emphasized **cultural values** asserting the importance of the individual, such as "individual personality" and "freedom." Within this historical context, it is difficult to open up to one's dependence by admitting shame.

Scheff's and Retzinger's studies alert us not only to one or two emotions but to the importance of our full range of emotions, and we can begin to make them more visible by labeling them with our range of concepts. For example, we can come to see the "seven deadly sins"—so labeled by Pope Gregory the Great at the turn of the sixth century—in a much different light with the aid of those concepts. Those "sins" are greed, lust, envy, sloth, gluttony, anger, and pride. We can come to see lust, gluttony, and greed as species of **addiction**. They all appear to be chronic, just as is the structural concept of addiction, and they are all very narrow or one-sided as well as outer-oriented, just as addiction constitutes subordination of individuality to dependence on external phenomena. Envy illustrates **relative deprivation:** both appear to be situational more than structural. Anger may be seen as a situational manifestation of a structure, **social stratification.** Sloth, perhaps similar to the withdrawal illustrated by James, may be an instance of **alienation.** As for pride, although biblical usages—"Pride goeth before destruction, and a haughty spirit before a fall" (Proverbs 16:18)—are negative, Scheff suggests that we label unjustified pride as "false pride," coming to see normal or justified pride in oneself as an **individual** or one's **personality structure** simply as pride.

Scheff's analysis and conclusions with respect to pride exemplify a very general approach not only to understanding problems but also to solving them, for what is involved is nothing less than changing the way we humans think, feel, and act from one moment to the next. If we think of pride or shame negatively, then those emotions will remain invisible yet will continue to affect us. We will have greater difficulty in being proud of ourselves for fear of appearing haughty and "falling," And we will continue to repress feelings of shame and as a result remain in "feeling traps" linked to patterns of **social stratification** and **alienation** along with an inability to develop egalitarian **social relationships,** and all of this will continue to be linked to our **bureaucratic worldview** and **culture.** However, concepts like those in boldface can help us all learn to become conscious of more and more of the complexity involved within any given human situation, following a key aspect of our web approach to the scientific method. This is not simply a question of developing such understanding with the aid of a psychotherapist, for it might be achieved as well through general education. If we turn once again to Figure 1-4, we might come to see the "expert" sociologist as in much the same boat as everyone else, just as Lewis's study revealed that therapists did not identify shame. If we sociologists, who are also Springdalers, are in fact to influence others to shift from a bureaucratic to an interactive worldview, then we must gain the pride required and become conscious of our hidden shame by first following the maxim, "Doctor, heal thyself."

PART III
SOME IMPLICATIONS FOR SOCIOLOGY

If we are indeed to build on Parts I and II, then what we have to build on is both extremely conservative and extremely radical. On the one hand we are returning to the Enlightenment dream for society, back to the visions of those French philosophers of the late eighteenth century, and back to Auguste Comte's nineteenth-century vision of a science of sociology. Anything seemed to be possible just prior to the French Revolution, and revolution was still in the air when Comte conceived of a new science of society that would capture the magical power of the physical and biological sciences. To return to those times we must somehow blot out what we are unable to blot out: the horrors of the twentieth century with its wars and Holocaust, its nuclear terror and environmental degradation, its inequality and addictions, its anomie and alienation, and its pessimism about the possibility of ever fulfilling the Enlightenment dream. On the other hand, those chapters suggest the realistic possibility of a new Enlightenment far exceeding the dreams of those French *philosophes,* a world where all of us can begin to fulfill the human capacities we have with the aid of a sociological imagination. In that world, where the twenty-first century might come to be designated as the age of social science, social technologies might come to be a match for technologies based on the physical and biological sciences.

Chapter 5 looks to the implications of the foregoing chapters from the perspective of the sociologist, centering on how we can actually move toward Alvin Gouldner's vision of a "reflexive sociology." We begin with a closer look at his argument in his *The Coming Crisis of Western Sociology.* Within the foregoing chapters the reflexive idea has been included as a key element of our reconstructed scientific method. Chapter 5's second section outlines possible new areas for substantive or basic research employing a web approach to the scientific method, by contrast with the more detailed examples of that approach in the foregoing chapters. Such substantive research is hardly ever reflexive, but it can become so when combined with an examination of investigator effect within those same studies. Next, we look to applied research.

Traditionally, such research centers on external groups. Here, however, our focus once again is reflexive, with the assumption that such research is desperately needed as a counterbalance to our **conformity** to the external emphasis within our **bureaucratic worldview.** Our focus is not on publishable work but rather on improving our understanding of self. Finally, we look specifically to our teaching activities, whether in higher education or elsewhere, again without stressing publication. For educational vision we turn to the ideas of Dewey, Freire, Illich, Pecotche, and Gandhi.

Chapter 6 is equally concerned with drawing out implications of the foregoing chapters for sociology, employing two sections: "Back to the Future," and "Forward to the Future." In the first, given the direction spelled out in Chapter 5 for a reflexive sociology, we are able to use that momentum—much like a pendulum—to gain further understanding of just how deep is the problem of changing our **cultural paradigm** and **worldview.** Here we add four concepts to help us achieve a wider perspective: **structure, situation, action,** and **interaction.** And we employ those concepts along with others to examine once again such topics as the invisibility of our problems, our usage of language in everyday life, and our repression of emotions. Yet we return to those earlier chapters in order to gain momentum to move more decisively into the future in our second section. To the extent that we sociologists have developed more comprehensive tools for understanding modern society than any other discipline—and I believe this to be true—then we need no longer be bashful about communicating what we have to offer. And if what we offer proves to be useful, then this will help us, reflexively, to learn just what our own potential is. Four examples where sociology might make its mark are debates about achieving a "civil society," directions for urban planning that emphasize improving opportunities for social interaction, the foreign policy of the United States, and the new social movements of the twentieth century, and we take up each of these areas.

At this point it is quite possible that many readers are wondering about the need for using boldface to emphasize the same set of concepts over and over again. Aren't we simply beating a dead horse, creating clichés and hitting the reader over the head when we should step back and allow the reader to develop his or her own sense as to what is and what is not important? I can understand such an attitude, since I too initially wondered about this usage. Yet I believe that we live in a world where it is perfectly fine to use technical concepts like "force," "mass," and "acceleration" for the physical and biological sciences, yet where any such usage for the social sciences is immediately **labeled** as a tiresome employment of clichés. I believe that this attitude is a small symptom of the one-sidedness of modern **culture,** which has come to accept rather than rebel against the failures of the social sciences along with the failures of our social technologies. I also believe, following the argument presented in Chapter 3—where bold-face was introduced—that the reader should attempt to see such usage as the tip of a vast iceberg, warning us of an invisible danger that we dare not face. That iceberg is a **bureaucratic cultural paradigm** and **worldview,** which teaches all of us to use *all of our words*—whether in thought, speech, or writing—in ways that pay little attention to the complexity of what we are attempting to understand and communicate. From this perspective, I suggest once again that when the reader encounters a word in boldface, she or he attempt to invoke the system of concepts within which that word is located.

5

Reflexivity

Following the sociology of knowledge, we are not simply dreaming in a vacuum about reflexivity, basic and applied research, and teaching. Our reconstructed scientific method suggests that we are now involved in an invisible crisis of mammoth proportions and that this crisis is rapidly accelerating. Granting the possibility that further research may disprove the validity of this hypothesis, enough knowledge from previous chapters has pointed in this direction so that we might do well to take it seriously. Even I, the author, have difficulty in accepting it fully, yet I am pushing myself to defer to existing sociological knowledge as pulled together within this reconstructed scientific method. From this perspective, our situation in modern society is an urgent one. And we sociologists in particular bear an enormous responsibility for learning about our situation and developing the tools essential to addressing it effectively. I believe that the time is long past when we might have succeeded in escaping from that responsibility, given the escalation of problems and the sparsity of solutions. Even if we believe that we have no interest whatsoever in problem-solving but are only interested in gaining substantive knowledge, the foregoing chapters indicate that the two are intimately linked. Our failure to be reflexive will yield, following those arguments, a corresponding failure to become aware of the complexities of human phenomena.

TOWARD A REFLEXIVE SOCIOLOGY

Gouldner's Vision

Alvin Gouldner's last chapter of *The Coming Crisis of Western Sociology* expresses his conception of a reflexive sociology in thirty-two pages, yet let us at least examine a few paragraphs to get the flavor of his thinking:

> Sociologists are no more ready than other men to cast a cold eye on their own doings. . . . Yet, first and foremost, a Reflexive Sociology is concerned with

what sociologists want to do and with what, in fact, they actually do in the world. . . . What sociologists now most require from a Reflexive Sociology, however, is not just one more specialization, not just another topic for panel meetings at professional conventions. . . . The historical mission of a Reflexive Sociology as I conceive it, however, would be to *transform* the sociologist, to penetrate deeply into his daily life and work, enriching them with new sensitivities, and to raise the sociologist's self-awareness to a new historical level. . . . A Reflexive Sociology means that we sociologists must—at the very least—acquire the ingrained *habit* of viewing our own beliefs as we now view those held by others . . .

In a knowing conceived as awareness, the concern is not with "discovering" the truth about a social world regarded as external to the knower, but with seeing truth as growing out of the knower's encounter with the world and his effort to order his experience with it. The knower's knowing of himself—of who, what, and where he is—on the one hand, and of others and their social worlds, on the other, are two sides of a single process. . . . The character and quality of such knowing is molded not only by a man's technical skills or even by his intelligence alone, but also by all that he is and wants, by his courage no less than his talent, by his passion no less than his objectivity. It depends on all that a man does and lives. In the last analysis, if a man wants to change what he knows he must change how he lives; he must change his *praxis* in the world . . .

The core of a Reflexive Sociology, then, is the attitude it fosters toward those parts of the social world *closest* to the sociologist—his own university, his own profession and its associations, his professional role, and importantly, his students, and himself—rather than toward only the remotest parts of his social surround. A Reflexive Sociology is distinguished by its refusal to segregate the intimate or personal from the public or collective, or the everyday life from the occasional "political" act. It rejects the old-style closed-office politics no less than the old-style public politics. A Reflexive Sociology is not a bundle of technical skills; it is a conception of how to live and a total praxis. (1970:487–90, 493, 504)

Gouldner's reference in the first paragraph to reflexive sociology's concern with "what sociologists want to do and with what, in fact, they actually do in the world" harks back to Figure 1-1, with its curves of aspiration and fulfillment and the gap between them. Gouldner moves from **social structure** to the structure of the **individual,** also implying—with his idea that we "cast a cold eye" at our own doings—that we should learn to perceive that same gap within our personal behavior. And to the extent that we learn to make a "habit" of such behavior or actually change our **personality structure,** then we can "transform" ourselves. We can draw an analogy between our own usage of the concept of **interactive worldview** and Gouldner's idea of a reflexive sociology. They contrast with a **bureaucratic worldview,** as illustrated by "just one more specialization," "just another topic for panel meetings at professional conventions," "concern . . . with 'discovering' the truth about a social world regarded as external to the knower," and "the attitude it fosters . . . toward only the remotest parts of his social surround." For Gouldner, what is necessarily involved for such a transformation to take place is the transformation of our own moment-to-moment activities in daily life: informal interactions with students, colleagues, family, and friends. In other words, "all that a man does and lives."

Gouldner's second paragraph may sound overly philosophical to many readers, yet it can be seen in a more mundane way as questioning all research that does not take into account "investigator effect" or the impact of the researcher on the respondent, those he or she observes, or the way the research comes to be interpreted. For example, has any researcher ever attempted to probe his or her own **worldview** and communicated this to the audience for the research report? To what extent has any of us even learned the nature of our own **worldview**? Later in that paragraph Gouldner suggests the importance of going beyond an understanding of our ideas in our quest for self-knowledge so as to take into account one's "courage" and one's "passion." Here we might look to the column under and including **cultural values** within social structure in Figure 1-3: **relative deprivation** and **reinforcement** within the situation, and **alienation** within the individual. What Gouldner says here about courage and passion is much the same as what Nietzsche maintained in his emphasis on the importance of emotions for the scientist in *The Gay Science* ([1887] 1974). Yet in addition to emotions and the intellect there is also the centrality of "praxis" in the world, getting at the column under **social organization** in Figure 1-3: **social stratification, bureaucracy, group, conformity, deviance, social interaction,** and **addiction.**

We may see Gouldner's vision of a reflexive sociology as an important first step in moving toward procedures that sociologists and all others can use to become aware of their **bureaucratic worldviews,** challenge them with the aid of these boldface concepts and move toward the **interactive worldview** that Gouldner describes as a "Reflexive Sociology." Following Gouldner's analysis above, that first step is a very large one. For one thing, we will have to look at our own personal failures to fulfill our ideals, and Gouldner himself fails to go very far in doing this in his omission of the conceptual tools of sociology—such as the boldface concepts—in discussing a reflexive sociology. In his response—quoted partially in Chapter 1—to a review of his book, he indicates awareness of language's centrality in moving toward a reflexive sociology:

The pursuit of hermeneutic understanding, however, cannot promise that men as we now find them, with their everyday language and understanding,

will always be capable of further understanding and of liberating themselves. At decisive points the ordinary language and conventional understandings fail and must be transcended. It is essentially the task of the social
sciences, more generally, to create new and "extraordinary" languages, to
help men learn to speak them, and to mediate between the deficient understandings of ordinary language and the different and liberating perspectives
of the extraordinary languages of social theory. . . . To say social theorists are
concept-creators means that they are not merely in the *knowledge*-creating
business, but also in the *language*-reform and language-creating business. In
other words, they are from the beginning involved in creating a new *culture*.
(1972:16)

Gouldner himself uses ordinary language in his discussion of reflexive
sociology, yet here he calls for usage of the "extraordinary" language of the
social sciences. This is exactly what we have attempted throughout this
book with our emphasis on the importance of sociological concepts at a
high level of abstraction and our introduction of twenty-eight of them
along with efforts to illustrate them. Of course, there is more to an effort to
reconstructing the scientific method than the use of abstract concepts coupled with a reflexive orientation where one applies those concepts to one's
own behavior. In addition, we must feel free to define problems that are
absolutely fundamental to self and world, and here Gouldner helps us as
well in his analysis. He sketches for us the potential impact of such an orientation on the individual, pointing the sociologist toward the importance
of emotions like "courage" and "passion." And reflexivity would also
point the individual toward "praxis" or effective actions in the world. A reflexive sociology would, then, fulfill the aims of critical sociology: to transform self and world. Gouldner's vision here is much like that of C. Wright
Mills's vision of the sociological imagination, which also was a direction
for sociologists and all others to learn to apply the concepts of sociology to
their own lives and as a result transform their understanding of self and
world.

Yet granting the importance of these visions, they share with all of sociology's special fields and all of academia the fact that they remain partial
visions. Despite the aspirations of sociologists, such visions fail to achieve
escape velocity from a **bureaucratic worldview.** We require in addition a
more complete understanding of a reconstructed scientific method. Gouldner and Mills start this with their view of fundamental problems in society
and the individual, their reflexive orientation, and their vague idea of our
learning to apply sociology to our everyday lives. Beyond that vagueness
we can bring to bear on our experiences the key concepts from the discipline. In addition, to complete our understanding of a reconstructed
scientific method we must come very far down language's ladder of abstraction, just as Mills wrote about shuttling far down as well as up that

ladder in *The Sociological Imagination*. And we must also learn to build bridges across the tower of Babel that is sociology and academia so that we have a platform of knowledge from the discipline and the social sciences as a whole that goes far beyond mere expert knowledge derived from this field or that one. It is then that we will be able to build the powerful social technologies on that platform—just as engineering builds on physics—to learn how all of us can move toward reflexivity or an **interactive worldview** and, along the way, confront our escalating social problems.

Yet let us be clear about the limitations of these boldface concepts given the nature of our present **bureaucratic worldview**. The latter centers on the intellect at the expense of the emotions, just as Simmel suggested many years ago. It is all too easy to ignore that imbalance and actually move to reinforce it as we proceed to learn to employ a web approach to the scientific method. This situation is much like that which often occurs in psychotherapy, where the patient talks about her or his emotions yet fails to *feel* the problems being discussed, and where the therapist must remain alert to the patient's repression of emotions. Each one of these boldface concepts should not only remind us of our system of boldface concepts but also of the deep personal and world *problems* that we are experiencing as a result of our bureaucratic worldview. Further, that reminder should in turn motivate us toward *taking action* to address those problems. Metaphorically, boldface should be a reminder not only of problems and potential solutions with our "head," but also with our "heart" and "hand." This carries forward Gouldner's broad vision. Yet to continue on such a journey it is essential that we make more use of our web approach to the scientific method, our orientation to emotional expression, and also our efforts at praxis.

Carrying Forward Gouldner's Vision

Illustrating further Gouldner's vision, we shall begin with substantive research on the nature of our **worldview**. This will be research that embodies a reconstructed scientific method. In the analysis of **relative deprivation** in Chapter 4 we saw that concept as a "missing link" that pulled the concept of **social stratification** into the momentary scene. Relative deprivation was defined as "the individual's feeling of unjustified loss or frustration of value fulfillment relative to others who are seen as enjoying greater fulfillment." With this definition, cultural values are pulled into the momentary scene no less than patterns of social stratification, thus bringing both **culture** and **social organization** into the situation. To achieve this with actual concepts that are tied systematically to a range of other concepts, and to apply it to a particular study, is a very far cry from vague talk about the importance of linking macrosociology with microsociology. Let us recall that Levin developed procedures for determining which individuals were closer to a **bureaucratic worldview** and which ones were closer to an **interactive worldview,** granting that they did not go very far in the latter direction. And we must also realize that it is our secondary analysis, and not Levin's study back in 1968, that is able to bring to the fore the study's implications for worldviews. More accurately, it is this tertiary analysis that is crucial.

Those closer to a bureaucratic worldview compared themselves with others, whereas those who differed somewhat from that worldview tended to compare themselves to their own previous performances. Here we should bear in mind the great emphasis of the bureaucratic worldview on the same outward orientation stressed so much within patterns of **social stratification, bureaucracy,** and **cultural conformity** that pervade modern society. Given this situation, anyone daring to compare self to one's own previous performances rather than to others is **deviating** from patterns of social organization, although at the same time **conforming** to key **cultural values** like "individual personality," "freedom," and "equality." And the result of such movement toward an interactive worldview appears to be escape from typical patterns of prejudice associated with patterns of **social stratification** and feelings of **relative deprivation.** Levin's research suggests, then, a direction that would move us toward an **interactive worldview,** where the individual would learn to gain **reinforcement** by fulfilling fundamental **cultural values.** Using the metaphor we used in Chapter 3, the individual would have the satisfaction of moving up a stairway with very wide steps, by contrast with the guilt, shame, and fear associated with acting so as to go against those cultural values.

This is of course no more than a tertiary analysis of one study. In order to learn a great deal about these two worldviews we require massive research in this direction, research within every one of our forty sections of the American Sociological Association. Assuming that reflexivity or an interactive worldview is crucial to uncovering ever more of the complexity of human phenomena, then all of our present research will necessarily remain limited until the human beings conducting that research learn to move into an interactive worldview. To illustrate further, there is no research without investigators performing that research, yet within our **bureaucratic worldview** almost no attention is paid to the phenomenon of investigator effect, just as Gouldner wrote that the knower's knowing of self and knowing of others are two parts of the same process. And even with the few attempts in this direction, they are severely limited by the knower's failure to know self. Yet even if the discipline as a whole centers on this kind of research, that in itself will not change the worldview of the sociologists involved in or reading the results of that research. In addition to such massive research, which I believe should be pursued urgently in view of what I see as the invisible crisis of modern society, I believe that we *also* require a focus on social technologies for helping us all to move ever more rapidly toward an **interactive worldview.**

We need not assume that it is premature to illustrate such technologies, since even misdirected efforts can prove to be valuable in developing more effective ones. I believe that not only is the very effectiveness as sociologists of every one of us at stake, but also that it is the sociologist more than anyone else at this time in history who is in a position to address our escalating problems. We might think of a number of stages that would be involved in moving from a **bureaucratic** to an **interactive worldview,** taking into account that our worldview is held in place by nothing less than our **cultural paradigm** and our patterns of **social organization.** First, we will have to learn how to apply the abstract concepts of the discipline—systematically linked with one another—to concrete external situations, such as those we experience in everyday life. Second, we will have to learn how to apply that system of concepts to momentary scenes that we experience personally. And third, we will have to employ the greater understanding and emotional development resulting from this education to what Gouldner has called "praxis": problem-solving in our own lives as well as externally. We can, thus, come to see movement toward Gouldner's reflexive sociology—or our own concept of an interactive worldview—as resulting from an effective social technology. These three stages correspond, metaphorically, to an emphasis on the "head," the "heart," and the "hand." And they could proceed in much the same way that we all have learned ordinary language, which we use as a tool to solve problems. As a result we could obtain **positive reinforcements** from such usage, thus creating our own procedures for learning to reinforce ourselves in more and more situations throughout our lives.

Before illustrating these technological stages, it is useful to step back for a moment and consider the nature of our assumptions. We assume that our reconstruction of the scientific method will prove useful in helping to integrate sociological knowledge, yet we have no more than begun to demonstrate that utility. Further, we assume that an interactive worldview—which that reconstructed methodology will help us to develop—will help us to understand and address social problems such as prejudice against minority groups. Finally, we assume that it is important to develop social technologies for helping sociologists and others learn to change from a bureaucratic to an interactive worldview. Shouldn't we, instead, wait for a more complete demonstration of the utility of this methodology before taking seriously its substantive implications? And shouldn't we wait until those implications are firmly established before moving on to develop social technologies based on those implications? Or should we instead follow one of Marx's *Theses on Feuerbach*, where he claimed that the point of philosophy is not to understand the world, but rather to change it?

In my own view, both of these either-or alternatives illustrate a **bureaucratic worldview.** A third alternative stems from an **interactive worldview,** where we do not have to choose between basic knowledge and the application of knowledge, just as we do not have to choose between basic so-ciology and applied sociology. If we fail to push ahead to develop a technology for helping the individual to change worldviews—granting the assumptions that have been only partially tested—then we will close off any opportunity to test the impact of an interactive worldview on the development of basic and applied knowledge within sociology. The result of that change in worldview might well prove to be the acceleration of substantive and applied knowledge. And such knowledge in turn might help us further in developing technologies to shift worldviews if that proves to be useful. In other words, we must change the world in order to understand it, *and* we must understand the world in order to change it. Even under the best of circumstances, we can never attain certainty or even the certainty that our knowledge will continue to increase. By moving ahead to examine social technologies for helping us shift to an interactive worldview we make no claim that the methodological and theoretical assumptions we are making have been proven useful or validated any more than partially. Yet by failing to move toward such technologies we may close off the real possibility of developing our scientific knowledge and learning how to understand and address society's fundamental social problems.

Social Technology for a Reflexive Sociology

At this point I will attempt to illustrate each of these stages in the development of an **interactive worldview.** I introduced a course, "Sociology through Film," which I taught for some fifteen years at Boston University. I would show short clips from classical films such as Fellini's *8 1/2* or Bunuel's *Exterminating Angel* to my classes, with the course involving substantial reading and discussion unrelated to those clips. We would then apply the abstract sociological concepts we learned in the course to the clips immediately after watching them. The three examinations in the course encompassed both our readings as well as essays that applied sociological concepts to those films. All of this has to do with the first stage outlined above: learning to apply a system of concepts to film scenes. Given the limitations of a college course meeting three times a week and having three examinations, this was no more than a beginning in learning how to use those concepts in everyday life. Although I supplemented our readings with manuscripts of my own, I myself had gone only a limited distance in understanding or presenting what we were trying to do in a highly systematic way. Yet despite these limitations, I believe that substantial progress was made in stage 1 for at least a minority of students and certainly for myself as well.

Here are two illustrations of the approach for stage 1 that I wrote for my classes. Each one centers on only one concept, although in our film discussions we invoked many concepts for any given scene we discussed:

Social Stratification (Fellini's 8 1/2)

Guido is lectured about what he must do. The cardinal, in his dream, tells him that happiness is unimportant and that salvation can only be achieved through the Church. After his dream, Guido asks the cardinal for advice about his film. Yet the cardinal pays little attention to Guido's question. And there sits Guido, listening to the cardinal just as he listened to his collaborator and his friends, who also told him how he should live his life. By looking up to the cardinal and to his friends, Guido is sacrificing himself: they are concerned with their own interests and not his.

Conformity (Bunuel's The Exterminating Angel)

In Luis Bunuel's film, one of the earliest scenes is that of the dinner guests who are seen entering the mansion and a few moments later seen entering once again, with the identical film footage and sound. This repetitiveness—reminiscent of Nietzsche's idea of "eternal recurrence"—is a theme for the film as a whole: the host's identical repetition of a toast, the continuing failure of the guests in their efforts to leave the room, the repeated inability of those on the outside to enter the house, the endless cycle of aggression, self-sacrifice, and hopelessness of the guests, and the repetition of a failure to be able to leave in the church scene.

These illustrations are based on the idea that audiovisual exercises might be transferred to experiences in everyday life. However, this is only one kind of example. There are a great many methods of learning how to speak or think in a particular language, such as audiotapes, classroom experiences, board games, electronic games, workshops, and computer software. If our bureaucratic worldview is as pervasive as it appears to be, then we continue to remain in its grip from one moment to the next within our thoughts, feelings, and actions. Under these circumstances, the individual requires the invention of procedures that can help him or her to move toward an interactive worldview in more and more situations. And the individual also requires more and more experiences that yield **reinforcement** of **cultural values** when making such movement. Of course, the invention of effective technologies depends on the basic understanding we have achieved as to the process that is involved. Just as such technologies

can help the sociologist or others to open up to the complexity of phenomena, so can our knowledge about how worldviews are changed help
us to develop effective technologies.

As for stage 2, or the application of abstract sociological concepts to one's
own everyday experiences in particular situations, there are a great many
ways of learning how to do this. Once again, it is essential that—just as
in the case of ordinary language—such learning yield **reinforcement** of
cultural values. One simple technique is the use of a diary, as illustrated
below:

> 5/10/00, 2:57 P.M. I'm sitting at the kitchen table revising my manuscript on
> the scientific method and thinking about . . . the fact that my desk is almost
> always cluttered. This illustrates the outer orientation of **addiction** and a **bu-
> reaucratic worldview.** I get my kicks out of something new rather than com-
> pleting something old. . . . There is a lack of the kind of **self-image** in which
> what is most important is within me rather than out there. . . . But while writ-
> ing this I am more at peace with myself, gaining **reinforcement** by looking
> back at my own behavior.

What is crucial here is that I confront contradictions between my own behavior and my basic **values,** that this helps me to become aware of those
contradictions, and that this helps me to resolve those contradictions over
a period of time. Of course, a few diary entries here and there are as nothing compared to my lack of awareness of inner contradictions in one scene
after another within everyday life. Nevertheless, those entries provide a direction, and social technologies might be developed to accelerate movement in that direction.

Stage 3 has to do with praxis, or actions that are sufficiently effective to
move toward fundamental individual and social change. And just as stage
2 is dependent on progress in stage 1, so is progress in stage 3 dependent
on accomplishments within the earlier stages. My illustration here is this
book, and the reader more than the author is in the best position to assess
the effectiveness of this "action." Of course, even if that assessment proves
to be positive, this suggests no more than an initial movement in the direction of an interactive worldview. Following Gouldner's analysis of reflexive sociology, such movement depends not simply on a certain kind of
knowledge, some particular development of one's emotional life, and
some specific kind of action. Rather, it depends on one's entire way of life,
on all of one's experiences. For the sociologist, for example, it will usually
depend—among other things—on how he or she goes about the business
of substantive research. In the next section we take up the implications of
our reconstructed scientific method for such research. Here, we do not assume that the sociologist has already changed worldviews, but simply that
he or she has begun to make that transition.

BASIC RESEARCH

In this section we begin by stepping back from our reconstructed scientific method in order to understand more fully just how it might be used by sociologists with many different orientations who are accustomed to using a wide range of different concepts. In previous chapters we have presented twenty-eight sociological concepts, yet literally hundreds are used within the discipline. Also, if this approach is one that may be utilized by other social scientists, then their usage of sociological concepts is minimal. Given this situation, we shall spell out a general approach to concepts that any sociologist and any other social scientist might choose to use. In our second section we examine some implications of this reconstructed scientific method for basic research within sociology. Our focus will be on the phenomenon of social change, granting that any other topic could be addressed as well. We shall not pursue our examples in any detail as has been done in earlier chapters. Instead, we shall make suggestions about a variety of areas for study. In each case the crucial question has to do with the bridges we are able to build connecting literatures that otherwise would remain apart. By using this reconstructed method are we able to develop directions for research that might lead to deeper insight into the phenomena under investigation? It is a question that can only be posed here and can only be answered by means of further research.

The Reconstructed Method as a Tool for All Social Scientists

If we return to Figure 1-3 we will note the twenty-six concepts depicted there and defined in the Glossary, with two concepts—social relationship and domination—added in Chapter 4. Throughout the foregoing chapters we have made use of this set of concepts, singling them out in boldface since Chapter 3, as tools to yield insight into the illustrations. They are all, with two or three exceptions, important concepts within the contemporary sociological literature, and their usage is based on their utility within a great many sociological studies. Further, they are located within a very broad framework encompassing culture, social organization, the individual, and the situation. And that framework also extends to include elements within the various structures and situations, such as thinking, feeling, and action within the individual, which enable us to understand change or process. That framework illustrates movement ever further up language's ladder of abstraction: from the abstract concepts to still more general concepts. And it is those very general concepts that can be used by any sociologist or any other social scientist as a conceptual guide to confronting the complexity and dynamism of human behavior.

Figure 5-1 presents this framework, as noted by the headings for the columns and rows, with illustrative concepts resulting from the cross-tabulation of those headings in parentheses within each numbered box. Those parenthetical concepts are no more than illustrations derived from Figure 1-3. If the full approach to dealing with complex phenomena is be invoked, then *some* concept should be used for each of the nine boxes. They can come from sociology or from other social sciences. In any case, they should have

ELEMENTS OF BEHAVIOR

	"Head": Beliefs, Ideas	"Heart": Interests, Aspirations	"Hand": Action, Interaction
Social Structures: Shared & Persisting Patterns	1 (anomie)	2 (cultural values)	3 (social stratification)
Situations: Momentary Behavior in a Scene	4 (label)	5 (relative deprivation)	6 (conformity)
Individual Structures: Persisting Behavior	7 (worldview)	8 (alienation)	9 (addiction)

CONSTRUCTION OF BEHAVIOR

Figure 5-1. Elements of behavior and construction of behavior.

emerged from some literature within a social science discipline so that they are able to carry the weight of that literature. It should be emphasized that none of the boxes is completely distinct from the others. Rather, each one suggests an *emphasis* on some phenomena versus others. For example, both **social stratification** and **relative deprivation** invoke hierarchies. But "social stratification" emphasizes a continuing hierarchical relationship, whereas "relative deprivation" centers on one situation. Also, "social stratification" emphasizes somewhat visible patterns of **social interaction,** whereas "relative deprivation" emphasizes relatively invisible **values.** Not all concepts fit neatly into these boxes, just as **culture** and **institution** and "role" (not one of our twenty-eight concepts) straddle boxes 1 and 2, but this does not alter the possibility of encompassing all of the boxes.

If we compare Figure 5-1 with Figure 1-3, we can see how much simpler the former is. Nine concepts rather than twenty-six are involved, and we are relatively free to choose our own concepts since what is most important is coverage of the range of very abstract concepts in the headings. If we try to make use of this table by applying it to our usage of any of the boldface concepts in this text, then this is a far simpler matter than attempting to see that boldface concept in relation to twenty-five other concepts. To use **culture** as an example, by moving across the column headings we can note that it encompasses, metaphorically, "head" and "heart" but not "hand." Thus, by centering only on culture we fail to take into account the important idea of social organization. As for the row headings, we might note that "culture" is an instance of social structure but fails to emphasize either the individual or the situation. And because of this lack, our understanding of cultural change will necessarily remain limited unless we link that concept to other concepts bearing on the individual and the situation. This approach is sufficiently simple so that it can be extended to our usage of any concept whatsoever within any situation whatsoever. By using the approach we move away from the simplistic orientation of our **bureaucratic worldview** and toward the complexity of an **interactive** one.

The utility of this framework remains to be fully tested, although it is suggested by the materials in the foregoing chapters. To repeat a basic example used throughout those materials and presented in Figure 1-1, it would not be possible to detect the existence of an "invisible crisis of modern society" without including both culture (within boxes 1 and 2) and social organization (within box 3). The crisis is seen more clearly when we move from social structure to structures within the individual which, largely as a result of **socialization,** illustrate such basic problems as **alienation** (box 8) and **addiction** (box 9). It is when we move from the top and bottom structural rows to the middle row dealing with the momentary scene, combining structural factors with momentary experience, that we can understand more clearly the dynamics of this problem or how it developed over time. Currently, the emphasis by social constructionists is on the momentary scene, all the while eschewing the "reification" of paying attention to social structures. Structuralists, by contrast, generally focus on patterns of social organization and sometimes patterns of culture, avoiding attention to the scene. By using concepts dealing with *both* social structures *and* the situation, we can utilize insights from both broad areas of the discipline. And those sociologists located within either orientation can learn to expand their interests.

If the researcher adopts a pragmatic approach, then it is unnecessary to jump from research using concepts from only one box in Figure 5-1 to research that includes concepts from all nine boxes. Development of more effective research should simply require the inclusion of concepts from at least one area in addition to what one customarily employs. Yet given our long-term commitment to the idea of *ceteris paribus*, where we assume that other things are equal, can we not gain important new understanding by working with only one of those nine areas? Following a bureaucratic research paradigm and **worldview,** the answer would unequivocally be yes. Yet if we assess the results of that paradigm for the cumulative development of sociology and the social sciences, then we must question that approach and look for a more fruitful one. Researchers working with concepts in each of those nine areas have already demonstrated the importance of those areas. Central to our discipline, following Snow's analysis in Chapter 2, are commitments to "relational connections" as well as "contextual embeddedness," both of which point us toward going beyond any single box. We might also note the centrality of the idea of interaction, examined in the first section of Chapter 4, not only for sociology but for the physical and biological sciences as well. *Ceteris paribus* may have proved useful for relatively simple physical phenomena, but it appears—analogous to middle-range theory—to point away from understanding complexity.

To illustrate just how far we have twisted and turned to rationalize our **bureaucratic sociological paradigm** and **worldview,** how can we possibly explain our failure to assess the investigator effects within almost every one of our investigations? If anyone knows about the impact of **social interaction, social stratification, cultural norms,** and **conformity** on an individual respondent or experimental subject, regardless of whether the interviewer or experimenter strives mightily for value neutrality, it is the sociologist. And if our entire legal system is sensitive to assessing the veracity of an interested party through rules of evidence and cross-examination, how can we fail to assess the forces within the researcher pushing for a certain conclusion and their impact on the respondent or the reader of the final report? Of course, we do have some rules aimed at avoiding bias, but they fail to go very far in addressing this question. As for the impact of our **worldview,** I have seen nothing in the sociological literature on this question.

On this latter point, Blalock has something relevant to say:

> Much more problematic, however, are disciplinary biases that result in the neglect of whole sets of factors as being outside the province of study or presumed to have negligible impacts. Here a more catholic or eclectic orientation may be encouraged so as to introduce a much wider range of explanatory variables that may have gone unnoticed even by individual investigators of differing ideological persuasions. (1984:35)

Procedures like assuming *ceteris paribus* when in fact other things are *not* equal are what Blalock has in mind. Yet although this advice is well-intentioned it cannot be followed by all of us who remain deeply in the grip of a **bureaucratic worldview, cultural paradigm,** and sociological paradigm that eschews breadth in favor of specialization-with-little-communication. And how are we to recognize such a bias, which reveals itself in our failure to investigate investigator effect as well as our more general failure to achieve rapid cumulative development within sociology and the social sciences? An **interactive worldview** and sociological paradigm would, by contrast, point toward a study of investigator effect within any given research project. Such a study would include every stage of the project: from its inception to measurement and data collection procedures on to the final report and subsequent interpretations. And it would be most sensitive to the researcher's **worldview.**

Some Programmatic Ideas for Using a Web
Approach: Social Change Illustrated

In the preceding chapters we have developed a number of illustrations, including some detailed ones, of how a reconstructed scientific method might yield insights that would otherwise remain undetected. This section, by contrast, will be programmatic. We shall limit ourselves to simply touching on a number of research topics and suggest directions for future research within this reconstructed framework. These topics will all be located within the general area of social and cultural change so that we can approach them all in a relatively systematic way. Also, the topic of social change is central to the next section of this chapter, where we take up applied sociology. In this way we will be able to see the relationships among topics within the broad area of social change, whereas normally we would tend to see those topics as disparate. For example, consider these ten topics: culture lag theory, the iron law of oligarchy, the Pygmalion effect, the process of secularization, the unanticipated consequences of purposive action, increasing divorce rates, the life cycle of the church, the self-fulfilling prophecy, the change in emphasis from production to consumption, and the new social movements of the twentieth century. We shall take a very brief theoretical look at these topics within the context of a reconstructed scientific method along with our earlier materials, suggesting their relationships to those materials.

If we are to follow a reconstructed scientific method, then it is essential that a reflexive approach be involved within each of the following areas of investigation. For example, whatever each of us learns about ourself as an individual should be brought into any research situation in which we are involved, for those factors are part and parcel of the entire mix that produces our findings. And the reader is entitled to see the full range of factors involved so as to be able to assess those results. We need not start from scratch, for we know that we are products of society no less than anyone else. To the extent, for example, that we are aware of the nature of the **cultural values** and patterns of **social stratification,** of the **anomie, alienation,** and **addiction,** of the **cultural conformity,** and of the **worldview** present in modern society, then we also know a great deal about ourselves. In addition, over time we will also have learned about unique aspects of ourselves, such as the nature of some of our **values** apart from cultural values. All of this knowledge is no less relevant to interpreting our results than whatever we learn about those we observe or read about, our respondents or our subjects. And we need not shy away from opening up to the possibility of our own biases, since it is the burial of such factors—versus revealing them—that would make our study less scientific.

Culture Lag Theory

Starting with William F. Ogburn's "culture lag" theory (1957, 1964), he distinguishes between material and nonmaterial culture and holds that the former elements tend to build on one another, just as using bronze is based on the use of stone, and the use of iron is based on using both stone and bronze. For him, nonmaterial elements of culture, such as religion, art, and law, tend to replace one another rather than accumulate. As a result, changes in material culture will occur at a faster rate than in nonmaterial culture, leading to a time lag in the latter's adaptation to the former. Yet it appears to be the greater simplicity of physical versus human phenomena that has yielded—following the revolution in the physical sciences—the acceleration of physical technologies known as the industrial and technological revolution. Figures 1-1 and 1-4 with their growing anomie, alienation, and addiction suggest our slow ability to adapt to those revolutions. However, Figure 1-5 suggests that this lag is by no means inevitable but is, rather, a product of our failure to achieve rapid development of the social sciences. To the extent that we can in fact achieve this, then we can expect that lag to be reversed, with changes in physical technologies following rather than leading changes in social technologies. And this would be accompanied by a change from the materialistic emphasis within modern society to a humanistic emphasis.

The Iron Law of Oligarchy

In his *Political Parties* (1949) Robert Michels developed the principle that every large organization, no matter how egalitarian its ideals, must establish a bureaucracy with a few leaders monopolizing power if it is to put its program into effect. It is the presumed inevitability of this occurrence that makes it an "iron law." Michels studied socialist parties and progressive unions in Germany between World Wars I and II, but he meant his iron law to apply to organizations and social movements in general. Yet is it in fact true that bureaucratic organizational means must necessarily displace organizational ideals? Joseph Gusfield's (1968) study of the Women's Christian Temperance Union (WCTU) indicates no such inevitability. When the sale of alcoholic beverages became legal, it chose to remain outside the mainstream of Protestant thought and thus sacrificed its potential strength. And in our analysis of Gandhi's practice of *satyagraha* in Chapter 4 we find a striking example of someone able to join his means to the end in view and somehow manage to teach others to do the same. Given our bureaucratic cultural paradigm and worldview, these examples are exceptions to what occurs in general. But the fact that there are exceptions suggests the lack of inevitability not only of Michels's iron law but also of our bureaucratic cultural paradigm and worldview.

The Pygmalion Effect

Robert Rosenthal and Lenore Jacobson's *Pygmalion in the Classroom* [1968; see also Rosenthal (1966) on experimenter effect] describes an experiment at an elementary school where students were given a little-known IQ test in the late spring supposedly designed to predict sudden "blooming" or "spurting" by students who previously had shown limited promise. At the beginning of the fall semester teachers were given a list of students who had been so designated—one-fifth of the student body—whereas in fact they were chosen using a table of random numbers. That "experimental group" gained significantly in their IQ scores over the next year in comparison to the "control group," the other students in the school, but only for grades 1 (15 points) and 2 (10 points). From this result we might infer changed social relationships between the teachers and students in the experimental group in which those labeled students received more reinforcements throughout the school year. They learned to conform, within their everyday social interactions with their teachers, to the teachers' normative expectations for better performance and, over time, were socialized so as to raise their own expectations or values for doing well and change their own self images. We might also assume less of a rigidly stratified structure within the lower grades prior to the experiment, so that the experiment came to be used as a basis for subsequent social stratification. This contrasts with the upper grades, where stratification had already taken place. Our analysis suggests the importance of more detailed research, which could measure the specific situational experiences of the students and teachers.

The Process of Secularization

Max Weber quotes John Wesley, the founder of Methodism, in *The Protestant Ethic and the Spirit of Capitalism*:

> I fear, wherever riches have increased, the essence of religion has decreased in the same proportion. . . . For religion must necessarily produce both industry and frugality, and these cannot but produce riches. But as riches increase, so will pride, anger, and love of the world in all its branches. How then is it possible that Methodism, that is, a religion of the heart, though it flourishes now as a green bay tree, should continue in this state? ([1905] 1958:175)

Wesley illustrates here the long-term process of secularization within modern society, linked to the accelerating triumph of the physical and biological sciences and their associated technologies along with the relative failure of the social sciences and their technologies to teach us how to deal with emotions like "pride, anger, and love" and emphasize a broader direction in society than materialistic **cultural values.** Yet if this trend of secularization is itself nested within a **bureaucratic cultural paradigm** within which religion and science have come to be defined as antithetical with little **social interaction** between them, then we can better understand this result. And this suggests studies of groups that have succeeded in linking religion to effective problem-solving, such as Alcoholics Anonymous and the great variety of support groups modeled on it.

The Unanticipated Consequences of Purposive Action

Robert Merton (1936) has pointed out the importance of the "unanticipated consequences of purposive action," and Hilmar Raushenbush (1969) has given us many historical illustrations of this. For example, after World War I the Allies sought to keep Germany weak by forcing it to accept sole guilt for the war and by collecting large indemnities. Largely as a result, Hitler was able to build up a resentful nationalism, destroy democratic institutions, and create a powerful war machine. As another example, Britain attempted to prevent Indian independence by violently suppressing the Gandhian movement. This stirred the British and Indian conscience to give greater support to the independence movement. In both cases there was a failure to pay attention to the importance of **cultural values** such as "equality," "freedom," "individual personality," "achievement and success," and "material comfort." And that failure in turn suggests the widespread existence of a **bureaucratic cultural paradigm** and **worldview** oriented to a very limited understanding of human phenomena. That limited understanding appears to pervade not only the ideas of people in general but also the ideas of the specialized "experts" who advised leaders of the allies after World War I as well as the British government during the Indian independence movement. Broader approaches to foreign policy require study (see, for example, Johnston and Sampson 1994).

Increasing Divorce Rate

Richard Farson attempts to explain divorce by making use of what has been learned about revolutions:

> Researchers have found that revolutions do not break out when conditions are at their worst but when the situation has begun to get better. . . . Historians call this the problem of rising expectations. . . . How can a good marriage fail? . . . . Because of the heightened expectations present in good marriages, they are often in greater jeopardy than bad ones. (1977:169–70)

Farson's emphasis on expectations within the family **group** can be broadened to society as a whole, as illustrated by the curve of rising expectations for fulfilling a wide range of **cultural values,** as depicted in Figure 1-1. From this perspective, it is not just people in "good marriages" but every married individual who is more prone to divorce, given the general revolution of rising expectations. And given this modern situation, this is indirect evidence for a host of other increasing problems in modern society, such as **alienation** and **addiction.** Studies based on the content analysis of written materials over long periods of time can be initiated to measure such changes in cultural values as well as **cultural norms.** And we can have indirect analyses as well, given our web approach, such as measurements of the degree to which emotions are increasingly suppressed, which we would expect to accompany the widening gap in Figure 1-1.

The Life Cycle of the Church

David Moberg (1962) has enumerated stages in "the life cycle of the church," which parallel Michels's iron law of oligarchy. They take a religious group from an idealistic and informal sect rebelling against a formal and often corrupt church through to a stage of formalization along with the loss of original ideals as the sect changes into a church. Moberg's analysis adds to that of Michels in that he gives much attention to what happens to emotions within this transition. Initially, sect leaders illustrate emotional spontaneity, as illustrated by the intense physical reactions of the Shakers and the "Holy Rollers." But as the sect becomes transformed into a church there comes to be less emphasis on emotional expression and more on ideas as a basis for action. If we measure an individual's commitment to a bureaucratic worldview, just as Levin did indirectly in his study of prejudice as outlined in Chapter 3, then we can learn the extent to which emotional expression versus repression is linked to that worldview. And we can also learn how expression and repression relate to an interactive worldview. And just as the Gandhian movement for Indian independence generally contradicted Michels's iron law of oligarchy, so might it be possible to develop religious organizations that did not move away from emotional spontaneity and expression, depending on the worldview of the members. And more generally, it might be possible to develop *any organization whatsoever* in this way.

The Self-Fulfilling Prophecy

Robert Merton introduced the idea of "the self-fulfilling prophecy" with the example of a run on a bank during the Great Depression (1949:180–81). A bank might actually have been in excellent financial shape, but someone just laid off from work might have started a rumor that people were withdrawing their money from the bank. Long lines of anxious depositors could follow and—given no federal insurance at the time—the bank could in fact face bankruptcy as a result. Merton defined a self-fulfilling prophecy as "a false definition of the situation evoking a new behavior which makes the originally false conception come true." Richard Henshel, attempting to follow up this idea, has opened up the complexity of what is actually involved so that we must go beyond one or two concepts to achieve understanding (1982a 1982b, 1990, 1993; Henshel and Johnston 1987). Within a bureaucratic cultural paradigm and worldview by contrast with an interactive cultural paradigm and worldview, we fail to pay attention to that complexity. Further, the former emphasizes accurate prediction as the hallmark of a genuine science. The latter, however, makes no such demand. Instead, with its web orientation it can yield over time ever-increasing understanding of any given situation. For example, we can embark on detailed research of any given scientific or lay prophecy to examine the nature of the situation, the social structures, the individuals and the momentary and long-term dynamics that are involved in the prophecy.

The Change in Emphasis from Production
to Consumption

Leo Lowenthal (1956) performed a content analysis of biographies in *The Saturday Evening Post* and *Collier's*, comparing those appearing between 1901–1914 with those appearing in 1940–1941. He found a decline in the percentage of biographies devoted to political, business, and professional life and an increase—from 16 to 55 percent—of those devoted to entertainers. He concluded that there has been a marked shift from "idols of production" to "idols of consumption" (ibid.). It is one thing to describe a change in **culture,** yet to understand such change we must bring to bear a number of factors suggested within an **interactive worldview,** and research might be directed toward examining such factors. For example, Lowenthal suggested that readers who despair of becoming extremely successful in politics or business can more easily identify with entertainment heroes who like or dislike tomato juice, golf, and highballs. This explanation implies the increasing gap between aspirations or **cultural values** and the fulfillment of those values based on such factors as **social stratification,** as depicted in Figure 1-1. Research might detect the extent to which the **individual** reader does in fact feel this way in general. It might also explore the reader's thoughts and feelings just after reading a biography, attempting to get at, for example, his or her **definition of the situation,** feelings of **relative deprivation,** and feelings of positive and negative **reinforcement.**

The New Social Movements
of the Twentieth Century

The civil rights movement, the peace movement, the women's movement, the ecology or green movement, and the gay movement all appear to illustrate protests against the repercussions of existing **social structures** within modern society. They alert us to the failure of contemporary society to help us all fulfill our nonmaterial **cultural values,** such as "equality" or "individual personality," which has to do with the very survival of the **individual.** In general there is very little contact among individuals within these various movements, yet their similarities can be understood by invoking a web approach to the scientific method. And those similarities themselves appear to be derived from the very nature of our **bureaucratic cultural paradigm** and **worldview.** From this perspective, partial efforts to deal with a given type of inequality become self-limiting in their results, for the problem is in fact far more fundamental. Such analyses can open up the depths of the social problems addressed by any given social movement and help us to understand what would in fact be required if social change is to occur. And those analyses could also help to educate participants within these movements so that they gain a clearer idea of what policies would be more effective ones, an education that could also help them with a range of personal problems that we all experience in modern society.

APPLIED RESEARCH

Up to now in this chapter and the book as a whole our focus has been on publishable research. That outer-oriented focus, even in the case of attempting to do publishable reflexive research, **conforms** to a **bureaucratic cultural paradigm** and **worldview.** Yet if we are to learn to become more reflexive and move toward an **interactive cultural paradigm** and **worldview,** then we must learn to balance that outer orientation with an inner one so that we can achieve interaction between the two. Applied research also almost invariably conforms to a bureaucratic worldview, with its focus on doing research about phenomena external to ourselves and then communicating the results of that research to others, whether in publications, reports, or in person. And of course that worldview guides our own behavior from moment to moment in everyday life. In a sense we are all like Ralph Ellison's *Invisible Man*: we are there but we are not seen by others as an important entity to take into account. Worse, we fail to see ourselves. How can we possibly embark on a reflexive sociology, given our enormous ignorance as to who we are as well as our failure to see our own significance within the research process?

A web approach to the scientific method requires that we learn a great deal about ourselves, regardless of whether any of it is ever published. In this section we shall focus on this general problem, and in the next we shall take up teaching in particular, whether within the context of formal education or elsewhere. The above sections of this chapter, granting their focus on publication, give us a start in this direction. For example, Toward a Reflexive Sociology suggests the utility of using audiovisual materials to learn how to apply the abstract concepts employed within a web approach. And that section also suggests the utility of a diary, where the individual can learn to apply those concepts to his or her own behavior. In the above section on basic research, there is the idea of taking into account information on one's own impact on every stage of the research process when drawing conclusions. That requires that we expand every research project so as to take into account this source of complexity, which has traditionally been almost completely neglected. Yet if we assume the pervasiveness of a **cultural paradigm** and **worldview,** all of this is no more than a beginning. How can we go far beyond this in our effort to develop a reflexive orientation and point toward developing an **interactive worldview?**

One approach is that we expand that diary so that it becomes not merely a slight adjunct to our written activities but that it becomes no less important than other writing we do. Although our tendency in any writing, following our **bureaucratic cultural paradigm,** would be to focus on external phenomena, an **interactive cultural paradigm** points us toward giving equal time to ourselves. Further, given what appears to be increasing emotional repression associated with a widening gap between aspirations and fulfillment, we would also do well to emphasize our feelings. Particularly important are those feeling that show us in a bad light if we were to **conform** to **cultural norms** and **values,** such as shame, hate, anger, fear, anxiety, guilt, envy, and prejudice. Also, it would be useful if we could make use of our understanding of social change. For example, Kuhn has suggested that the scientist's recognition of contradictions within a scientific paradigm can lead to change in that paradigm if an alternative one that promises to resolve those contradictions is available. Studies such as those by Milton Rokeach and his associates (see, for example, Rokeach 1968; Ball-Rokeach 1984; and Greenstein 1989) have traced experimentally the impact of "self-confrontation" on the modification of beliefs as well as behavior. Here, we can combine self-confrontation with the alternative paradigms sketched here.

A diary can also be expanded so as to include events in one's past. We embody an ocean of past experiences that have shaped us, yet very few of us go back to memories of those experiences in an effort to understand them more fully. Even if those memories do not accurately reflect what in fact happened, our **definition of the situation** throughout our past experiences has served to shape our behavior. Working with a web approach to the scientific method we are in a position to understand more fully what happened, including our emotional reactions. In addition, when we have continued with a diary over a long period of time, those past entries give us an excellent opportunity to look back at ourselves, for we have the exact wording of those entries. The following example is taken from a diary entry I wrote on September 19, 1982:

> I got angry at someone's attitude toward me and then felt very small. It appears that anger can make me feel guilty. After a while I began to realize the foolishness of the anger and I accepted the person, felt less angry and guilty. Then I felt much better about myself. It appears that if I am to retain an extremely good image of myself, it is essential that I *work through* my *anger* as well as my guilt: (1) accept it as being OK under the circumstances, and (2) go beyond it: guilt *and* confidence, anger *and* acceptance, love.

Let us view this pragmatically: not as a full-blown analysis so thorough that it is worthy of publication but simply as an analysis that yields a bit more insight into one's behavior. Despite the talk about emotions here, it appears—in common with the **bureaucratic worldview**—that my dominant orientation was to repress certain emotions so that I could feel "extremely" good about myself. For example, my anger was immediately coupled with my feelings of guilt, and I can understand this from the nature of our **cultural norms:** we are not supposed to get angry but rather should remain "in control of"—read "in a state of emotional suppression of"—our anger. Perhaps, though, we have an excellent right to get angry, very angry, at a **cultural paradigm** that teaches us to repress our emotions. Of course, if we're not aware of the existence of such a paradigm then we will either take that anger out on others—who are equally victims of that paradigm—or on ourselves through such feelings as guilt. And since others can respond in kind, we might then become the victims of their rage. Overall, this entry reflects not only emotional repression but also a desire to accept the existing social relationship, rather than open up to the possibility of changing it. Here again we have an instance of the **bureaucratic worldview** with its continuing patterns of **social stratification** and **bureaucracy.** By this reanalysis, however, I begin to open up problems that I had closed down beforehand.

Here is an entry dated August 16, 1983, where I go back to an event I had experienced as a child:

> Scene: An argument between Mom and Dad in the kitchen where Dad shouted angrily and broke plates, I was terrified, and Mom cried. She confided in me that she was always able to bring my father around over a period of time. Whereas he would act out, she would sulk. Emotionally, she seemed to hold herself in tightly. She hated direct conflict, raising her voice. Her ideal was to be "nice," not to confront anyone. She was the youngest in her family, the baby. Emotionally, I never noted much warmth between her and Dad. She seemed to have a kind of servant role, catering to his needs. Dad was set up as "bad," Mom as "good."

Looking back at this entry in relation to the above-quoted one, I can understand a bit better my own emphasis on quickly getting rid of anger. It is not simply a case of emotional repression within society as a whole but **also** a case of my own **social relationships** within my own family. Apparently my mother suppressed her own emotions rather than confront my father, yet she was able to attain a **stratified** relationship of **dominance** by means of manipulation rather than confrontation. And apparently I have a great deal of work to do in order to open up to my own buried emotions.

In addition to a diary, there is the effort to use sociological concepts within more and more of the everyday scenes that we experience. Our problem is not that of applying twenty-eight concepts to every scene we experience but simply that of improving on our range of coverage of the nine categories in Figure 5-1. Further, given our understanding of the emotional problems within modern society in general, we might do well to focus some attention insofar as possible on categories 2, 5, and 8, which deal most directly with our emotions. To illustrate, to what extent do I stereotype my father as "bad," remaining unaware of his commitment to **cultural values** that I shared with him, and also remaining unaware that I also shared with him his patterns of **social stratification**? And to what extent do I stereotype my mother as "good," remaining unaware of what she did to achieve **dominance** over my father in a manipulative as distinct from overt way? To what extent do I **label** myself as someone who has solved such basic problems as **alienation,** failing to recognize the tremendous forces within modern society that push us all in this direction? Do I have difficulty expressing emotions because of a very wide gap between my own aspirations and their fulfillment?

In addition to attempting to pay attention to both **culture** and **social organization,** movement into an **interactive worldview** requires that we take into account the situation. When being passed or passing someone while driving, to what extent do we feel **relative deprivation** in the former case and **reinforcement** in the latter? And if we become aware of such feelings, to what extent do we feel shame or guilt for having them, thus moving toward feelings of **alienation**? Following an **interactive cultural paradigm,** can we learn to feel, instead, anger at a cultural paradigm that pushes us into patterns of **dominance** and subordination? Taking up another driving experience, to what extent do we drive faster when someone is tailgating us, feeling guilty or ashamed because of our relative slowness? And if we become aware of such **conformity,** do we then become ashamed of being conformists? Alternatively, do we get angry at the tailgater? If so and if we become aware of that anger, do we then feel ashamed or guilty about getting angry? Following our analysis of Scheff's and Retzinger's analyses in Chapter 4, do we frequently experience cycles of shame and rage or anger?

If such analyses of our everyday lives fail to produce positive **reinforcements** as well as more effective actions, then we will abandon them after a time no matter how hard we continue to try. For example, to the extent that one has become committed to moving toward an **interactive worldview,** then any recognition of movement beyond stereotyping, such as using one of these boldface concepts, will come to be seen as a positive **reinforcement.** For example, if we succeed in shifting from a feeling of **relative deprivation** when someone passes or attempts **domination** to a feeling of **social interaction** by seeing that other individual as much like us, then that too can yield positive **reinforcement.** Given the depth and pervasiveness of our **bureaucratic worldview,** however, we should not expect any rapid changes in our behavior. In looking back over the diary I kept for some fourteen months beginning in the fall of 1982, I am surprised at how little I have developed emotionally since that time. What we all appear to require if we are indeed to shift our **cultural paradigm** and **worldview** as well as the scientific paradigm nested within it is more than such efforts, granted that they move us in the desired direction. As sociologists we should be well aware of the power of the **group,** even if it is a small group. The group can become an important factor in this kind of change.

The analogies I will use here are women's consciousness-raising groups that flourished in the United States during the 1970s as well as support groups of all kinds that flourished then and still do. In both cases it is crucial that the group remain small enough and informal enough so that a good deal of egalitarian **social interaction** can take place, by contrast with the **social stratification** emphasized within formal groups. For example, the twelve-step groups modeled after Alcoholics Anonymous generally exclude leadership on the basis of professional credentials, and they have succeeded in attracting members with a very wide range of problems. The thirty or so such national groups are illustrated by Batterers Anonymous, Debtors Anonymous, Depressives Anonymous, Divorce Anonymous, Drugs Anonymous, Emotions Anonymous, Emotional Health Anonymous, Emphysema Anonymous, Families Anonymous, and Fundamentalists Anonymous. The popularity of such groups may be explained in part by their partial movement toward an **interactive cultural paradigm** and **worldview.** Assuming that the **bureaucratic cultural paradigm** and **worldview** not only lives but has succeeded in dominating society, then we are all quite outer-oriented and group-oriented. To the extent that even a small group within which we are involved adopts an alternative **cultural paradigm** and **worldview,** the **positive reinforcement** coming to us from such membership will be great.

Yet another approach to learning about ourselves has to do with technologies specifically invented for this purpose. These were briefly illustrated, largely with the aid of materials from my own course on sociology through film, in the above section, Toward a Reflexive Sociology. There a number of possible procedures were mentioned, such as the invention of board games, electronic games, workshops, and software developed particularly with this purpose in mind. To the extent that there does develop a growing movement pointing toward the development of an **interactive worldview,** such technologies can come to be analogous to the industrial revolution in bringing the economic institution into the arena. Just as technologies based on the physical and biological sciences have succeeded in moving our world in a materialistic direction, so might social technologies succeed in moving our world in a nonmaterialistic direction. Yet that movement need not eschew material development but merely shift priorities so that they are more balanced as between the two directions. What is essential is that it would come to be seen as potentially profitable to invest substantial sums in the development of such technologies. Given that the market for those technologies could ultimately include everyone on the planet, initial successes by a small group of sociologists could yield dramatic and long-term economic support.

If the web approach to the scientific method does indeed prove to be fruitful, then it should achieve momentum throughout society with little need for preaching, since market mechanisms would carry it forward. However, it is absolutely crucial that there be an initial demonstration of fruitfulness to the initial group that is involved, a demonstration sufficiently effective so as to stimulate the group to accelerate its own momentum. It would have to be a demonstration not merely of the effectiveness of a methodology for coming up with insights about external phenomena. In addition, it would have to help the group fulfill fundamental values encompassing the range of life's experiences. Given the power and comprehensiveness of our present **cultural paradigm** and **worldview,** we cannot be certain that such a change will take place. An escape velocity must somehow be reached. Reaching that velocity is not just a question of the acceleration of social problems within all institutions, for we Springdalers have learned very well how to bury those problems in the interest of gaining some degree of satisfaction with our lives. In addition, it is a question of developing an alternative **cultural paradigm** and **worldview** that promises to resolve those contradictions. And even more specifically, this depends on the fruitfulness of a reconstructed scientific method based on that alternative cultural paradigm.

TEACHING

This final section of the chapter is also oriented to achieving reflexivity or, more generally, movement toward an **interactive worldview.** Thus, our focus once again is not on research for purposes of publication but rather on learning how to teach as well as how to learn, and here a personal diary of one's teaching experience—which I never kept—may prove useful. There is certainly an important place for both basic and applied research on education, and we certainly need published knowledge in these areas. But this entire book points to the centrality of an interactive worldview for opening up to a scientific method that can in fact achieve rapid cumulative development in these and other areas. We begin this section with a negative illustration drawn from outside the context of formal education. We then draw on the ideas of several educators—John Dewey, Ivan Illich, Paulo Freire, and Carlos Pecotche—together with the approach adopted by Mohandas Gandhi. Finally, we turn to the idea of "reflexive teaching." My approach is to see such teaching as no more than the application of an interactive worldview to teaching. There, I briefly attempt to bring to bear aspects of the work of Dewey, Illich, Freire, Pecotche, and Gandhi to the teaching situation and to some reflections on my own teaching experiences.

A Negative Example: Consultants Using System Dynamics

We may view consulting as a form of teaching, but it rarely follows the kind of egalitarian interaction that I believe is central to effective teaching and learning. Following our **bureaucratic cultural paradigm** and in common with almost all education throughout the world, its emphasis is very heavily on one-way communication. As for "system dynamics," this is an approach to computer simulation developed by Jay Forrester and several colleagues at MIT's Sloan School of Management. We cited references to the work of this group after presenting Figure 1-4, since that figure's feedback-loop approach is a key idea within system dynamics. However, those working with system dynamics see such causal-loop diagrams as no more than an initial step in working toward a full-fledged computer simulation. The example to be presented is based on a paper I published in the proceedings of a conference on cybernetics and society with the title, "Paradigmatic Barriers to System Dynamics" (1980). As the reader may note, I was taken with Kuhn's analysis of scientific revolutions many years ago, and I still believe that this analysis can help us to change both sociology's scientific paradigm and the cultural paradigm of modern society. This example is not meant to single out system dynamics or consulting for special criticism. Rather, this is an example of basic problems involved in all teaching.

I begin by quoting from my own article:

> Let us, then, look for contradictions within SD that are linked to cultural anomalies. . . . One of SD's central ideas is that it is useful to conceive of societal phenomena as feedback or closed systems in contrast to open systems. . . . This way of seeing phenomena is associated with a second fundamental idea of SD: the importance of examining the long-term dynamic behavior of systems. Decisions that may seem perfectly rational to a policy-maker may actually intensify an initial problem, whereas counterintuitive behavior may solve it.
>
> Yet suppose that these two ways in which the system dynamicist perceives phenomena are no more than drops in his or her ocean of perception, and that the rest of the ocean is oriented to seeing open systems and very short-term behavior. Then that ocean of perception will seriously limit the system dynamicist's ability to carry SD thinking very far. Otherwise, one would have to believe that modeling activities can somehow remain isolated from the rest of a modeler's orientations, a belief contradicted by the SD worldview emphasizing closed systems. (1980:682)

Here I raise the same question for system dynamicists that I have raised in this book for sociologists: Do they/we preach one thing yet practice another? In their case it is preaching feedback loops yet practicing nonfeedback teaching or consulting. In our case it is preaching scientific ideals of openness to information yet practicing an interpretation of the scientific method that is relatively closed.

In the remainder of the article I attempt to spell out the nature of those "paradigmatic barriers to system dynamics." As a consultant or teacher, the "modelers share a basically managerial world view" (Meadows 1980:25), a quote from one of the leading system dynamicists. The modeler or teacher of system dynamics is the producer and the client or student is the consumer, or the modeler is the expert and the client or student is the layman. Thus, the basic relationship between modeler and client or student is an open system centering on short-term behavior. Yet far beyond this, and I base this conclusion partly on my own involvement with the system dynamics group, system dynamicists are largely ignorant of the social science literature, yet they fail to see the limitations of that ignorance. This conclusion can be easily checked by perusing the range of references to the social science literatures in their key books. For example, of the *eleven* references Forrester cites in his *World Dynamics* (1971), seven are to himself or his colleagues and three are to someone he is collaborating with in a research project. As a result, we have in system dynamics—and appear to have in teaching generally—good intentions coupled with contradictions, just as is illustrated in our Figure 1-1 with its increasing gap between aspirations and performance.

We can use this negative illustration to better understand a direction for how to join means and ends or situational behavior and structural goals such as Gandhi illustrated, an approach that proved to be enormously effective in achieving large-scale social change. For example, system dynamicists might teach their clients how to model, since the procedures involved are not that complex although they do take some time and effort. To the extent that the client's models prove useful, that will create a market for other consulting or teaching jobs. Far more difficult, however, is the task for the system dynamicist of opening up to his or her learning from the client or student. This parallels the failure of system dynamicists to open up to their limited knowledge of the social sciences. Meadows et al. revealed that limitation in their most well-known publication, *The Limits to Growth* (1972), which received extensive criticism because of assumptions not justified by the relevant literatures. The problems here are similar to those existing within sociology. How often do teachers believe that they can learn a great deal from their students? And how often do they believe that it is vital for them to go beyond their own specialized areas? In both cases we are dealing with nothing less than the **bureaucratic cultural**

paradigm and **worldview.** The teacher or modeler envisions no stairway for personal development based on an **interactive worldview.**

Dewey, Freire, Illich, Pecotche, and Gandhi

Let us look very briefly at key ideas from a number of educators along with some of Gandhi's orientations.

John Dewey

Dewey's approach was very scientific. He saw education as setting up problems to resolve rather than as providing solutions, and he rejected the "solutions" of most progressive educators. Dewey's concept of progressive education emphasized the cultivation of individuality, spontaneous student activities, learning through experience, and the importance of understanding the changing world. He emphasized the importance of emotions, since his is a problem-oriented approach where student spontaneity is crucial. For Dewey, ideas are tools for solving problems, but they should not overshadow both emotions and experience. His idea of "reconstructing philosophy"—as presented in Chapter 2—has much in common with our own approach to reconstructing the scientific method. And he saw the failure of the social sciences by contrast with the natural sciences as lying at the heart of the fundamental problems of modern society. Dewey points away from a **bureaucratic cultural paradigm** and toward an **interactive cultural paradigm.** Yet like C. Wright Mills and Alvin Gouldner, he is able to chart no systematic and specific path for moving very far in that direction.

Paulo Freire

Paulo Freire, born into a middle-class family in northern Brazil, suffered severely as a result of the economic problems of the Great Depression and vowed to fight against poverty. His *Pedagogy of the Oppressed* (1970) and *Education for Critical Consciousness* (1973) illustrate an approach to educating illiterate peasants that centers on the power of language and aims to transform them into people who can say, "I now realize I am a man, an educated man." For example, a small group of educators go into a given area, engage in **social interaction** with the agricultural workers there, and learn about the vocabulary they use and the problems they face, such as "slum," "plow," "wealth," "food," "sugar mill," "salary," and "government." They then use these words along with familiar images in small discussion groups or "**culture** circles," helping those involved to see how these words can help them to "transform the world." Their overall aim is to help people who have been submerged all of their lives within a "culture of silence," a culture characterized by a lack of self-expression in thought and speech. Freire's emphasis on language, on **social interaction,** on the problems experienced by the **individual** as well as his transformation, on **culture,** and on social change provides lessons for modern society, where people's "right to say their own word has been stolen from them, and that few things are more important than the struggle to win it back" (1970:15).

Ivan Illich

Ivan Illich, a defrocked priest who taught for many years in Latin America, questioned modern society's reliance on "institutional treatment" as rendering "independent accomplishment suspect": "Not only education but social reality itself has become schooled" (1972:3). Illich is concerned about the fact that we have all learned to be dependent on **institutions** such as schools and are therefore neglecting our own **individual** development, placing ourselves at the bottom of an institutional hierarchy. As a result, we learn "the hidden curriculum" of the school or of any other institution: that we are incompetent to learn, feel, or do on our own and must depend on authorities or experts to teach, feel, and do for us. . . . The student then fails to question the schools because they have been equated with education, just as the church has been equated in the minds of many with the path to salvation. Illich succeeds in using the concepts of "hierarchy," "schooling," "dependence," and "institution" in a very abstract way so that they can apply to all of society, just as we use the concepts of **social stratification, institution, dominance,** and **alienation** within our own web approach to the scientific method. By combining Freire's and Illich's ideas, we come up with the importance of sociology's abstract concepts in helping all of us to "say our own word" and escape our "culture of silence" so as to "transform the world."

Carlos Bernardo Gonzalez Pecotche (Raumsol)

Carlos Pecotche was born in Buenos Aires in 1901 and—like his fellow Latin Americans Freire and Illich—relied on powerful metaphors throughout his life (see, for example, 1985, 1986, 1991). In 1930 he founded the "School of Logosophy" in Cordoba, Argentina, centering on helping the individual learn to achieve "conscious evolution," with schools of logosophy later becoming established in Latin America and other countries. He devoted much attention to the power of the individual's momentary thoughts in "enslaving" each of us to such patterns of thought, feeling, and action as "greed," "haughtiness," "hypocrisy," "impatience," "inconstancy," "indifference," "indolence," "inhibition," "intolerance," "lack of will power," "rancor," "rigidity," "sullenness," "vanity," and "verbosity." He retained and communicated a vision of the incredible potential that each individual has to conquer such momentary thoughts and reconstruct his mind so as to evolve throughout his life. In this way, he saw the individual as developing alternative patterns of thought, feeling and action, such as "knowledge," "judiciousness," "sincerity," "honesty," "veracity," "perseverance," "interest," "self-determination," "resolution," "tolerance," "diligence," and "fortitude." Pecotche succeeded in pointing up the relevance of the individual's situational behavior—such as definition of the situation, label, relative deprivation, reinforcement, conformity, deviance, and social interaction—for the structure of the individual and society.

Mohandas Gandhi

Gandhi did not see himself primarily as an educator nor did others see him in that way, yet he embodied directions for education that are of central importance if we are to learn to move toward a reflexive sociology and an **interactive worldview.** Our analysis in Chapter 4 emphasized the unity of "means and ends" within his life, and we can also read this as the unity of situational and structural concepts. He was able to follow through on his egalitarian philosophy with emotions like trust and empathy and with actions like refraining from violence or humiliating the rival group and achieving friendly verbal discussions. His worldview as an end was part and parcel of his **social interaction** as a means. And visually we could see this man in little more than a loincloth as someone who practiced what he preached. We have of course learned subsequently that he was no saint in his treatment of his women companions, yet we have no hero from the past who was perfect by modern standards. Overall, Gandhi suggests the possibility that we moderns can in fact turn away from **alienation** and **social stratification** and learn to move toward the **cultural value** of equality and toward egalitarian **social interaction.** He suggests, then, the possibility that a teacher of students can in fact become someone who learns from those students, and that a student can in fact become a teacher of teachers.

Toward Reflexive Teaching

I conceive of "reflexive teaching" as no more than the application of an **interactive cultural paradigm** and **worldview,** coupled with a web approach to the scientific method, to teaching. This might be best understood by invoking some of the ideas of Dewey, Freire, Illich, Pecotche, and Gandhi. We can take Dewey's focus on problems to help a group, including the teacher, work together on such problems as confronting their own feelings of shame, guilt, fear, **alienation,** and **addiction.** From Freire the group can learn to say its own word, learning to use abstract concepts from the social sciences that extend over the nine categories of Figure 5-1. From Illich the group can come to question its own patterns of **social stratifica-**

tion and **conformity,** where they have adopted society's hierarchical way of life or **bureaucratic worldview** as a result of **socialization** or "schooling" within all **institutions.** From Pecotche the group can become oriented to the power of the momentary scene, as illustrated by thoughts as to their own inadequacy, in preventing their **individual** evolutionary development. And from Gandhi the group can learn to develop profound emotional commitment to confronting fundamental problems within society as well as their own lives, and also to act on the basis of that commitment.

Yet how is all this to be achieved, given our **bureaucratic cultural paradigm** and **worldview,** our **anomie, alienation,** and **addiction,** our patterns of **social stratification, bureaucracy,** and **cultural conformity,** and our bureaucratic interpretation of the scientific method? If we follow Kuhn, Dewey, Freire, Pecotche, Illich, and Gandhi then we can come to define the teaching experience—along with each of our experiences in life—as invoking the fundamental problems of human society. And we can also learn to follow them in developing the conviction that each of us possesses the incredible potential to learn to "say his own word" and shape society and self so that we all can evolve—in our thoughts, feelings, and actions—as individuals. Looking back at my own teaching experiences, primarily at Boston University, a key barrier to my own moving in this direction was the conviction that I was already in possession of the crucial answers, and that my task was to communicate those answers to my students. From my present perspective I no longer blame myself for my inadequacies, nor do I blame society. Rather, I see myself as having journeyed a certain distance toward reflexive teaching, for example, by using film clips and news stories to learn abstract concepts that moved myself and some others to examine emotional problems. From the perspective of an **interactive worldview,** this journey is an endless one. Yet if we have experienced the acceleration of social problems within modern society—assuming that human behavior is at the bottom of this—then we should also be able to experience the acceleration of solutions to those problems.

Also from this perspective, the problem of learning to teach reflexively can never be solved within the context of teaching alone, for that context is only a portion of the teacher's life. Can we imagine Gandhi accomplishing what he did by dividing his life into *satyagraha* activities and other activities? The ability to teach reflexively is nested within the ability to live reflexively, following Gouldner's vision. As for the latter, the development of that ability remains a problem for all of us. To the degree that we continue to see that as a problem, I believe that we will have made real progress in the area of emotional expression. I recall in my analysis of my diary, for example, how easy it was for me to pretend that I had solved the problem of emotional repression when in fact I had only just begun to face that problem. I wanted desperately to feel "extremely" good about my

self-image, rather than to learn that *the very uncovering of a problem* could constitute a **positive** Following Dewey's emphasis, life—and not just education—can become a process within which we are continuing to uncover problems, a process where we can learn to use the full power of the scientific method to confront those problems. And following Pecotche's orientation, we can learn to use such experiences as a basis for our own **individual** evolution together with the evolution of society.

6

Language and Emotions

If Chapter 5 emphasizes reflexive implications of a web interpretation of the scientific method, then Chapter 6 looks outward to external implications. Yet at this time we can do no more than hint at possibilities. Could anyone possibly have predicted, at the beginning of the seventeenth century, the impact of the physical and biological sciences and their associated technologies over the next four centuries? Nevertheless, our experiences up to the present do point in certain directions. We have learned over these four centuries the incredible power of both the scientific method and the technologies—associated with the physical and biological sciences—based on that method. For example, President Clinton's phrase, "It's the economy, stupid!" alerts us to the repercussions of a worldwide economy based on those technologies. This suggests that if we sociologists wish to have our voices heard in modern society we cannot afford to focus only on developing substantive knowledge. Yet neither can we afford to center solely on applied knowledge. The interactive approach to the scientific method and to our cultural paradigm and worldview sketched in the foregoing pages demands that we find the motivation and the time to emphasize *both* substantive and applied sociology. Although we emphasize the latter in this chapter, the former is no less vital for sociology and society.

We have already begun to explore some of the implications of the web approach to the scientific method along with an interactive worldview and cultural paradigm for social technology. In Figure 1-5, for example, we sketched a feedback diagram that could help us to understand how to close the gap between (rising) expectations and their fulfillment. Also, in our analysis of Gandhi's *satyagraha* we attempted to carry further our understanding of conflict resolution by emphasizing the importance of attention to *both* structures and the situation. Further, our analysis of emotions applied that same structural-situational approach to a particular case study and pointed a direction for becoming aware of and changing a range of emotions. Chapter 5 suggests technological procedures—such as using audiovisual materials and diaries—for learning to use sociological concepts within everyday life. And that chapter also attempts to pull together ideas

195

about effective teaching, building on the ideas of several approaches. All of these specific examples of social technology help to carry forward our understanding of how to proceed in this direction. Yet they lack an abstract or general underpinning, namely, their implications for an overall approach to language that is tied to our interactive orientation. It is language that has enabled us humans to become the most interactive creatures in the known universe, and it is to language in general that we must turn for a deeper understanding of social technology.

Paraphrasing President Clinton's phrase, we can claim, "It's language, stupid!" How could we humans have been so dense as not to understand the centrality of language as our most powerful tool for shaping ourselves and society? And how could we sociologists have been so blind, despite all that has been learned about language in many disciplines throughout the twentieth century, as not to see this handwriting on the wall of our knowledge? If we look back at the preceding chapters from this perspective, we can come to see that handwriting very clearly. For example, there is our focus on learning to use abstract sociological concepts within everyday life—carried further with our use of boldface—as a basis for solving substantive and applied problems. There is in Figure 5-1 a very general approach to language that anyone can learn to use without employing any of sociology's technical concepts. It is that approach that illustrates the systematic usage of language to be found within our web orientation to the scientific method as well as an interactive worldview and cultural paradigm. There is also our use of images in Figures 1-1 through 1-5 and 4-1 which point up the importance of perception for us organisms. And there is in addition our freedom to embark on using figurative language like metaphor and simile, in common with the freedom of the poet or novelist. Also, there is our reflexive perspective, for—whatever else we are or come to be—we humans are creatures of language.

In this chapter we continue to use boldface on occasion for the thirty-nine concepts appearing in Figure 6-1. Granting that we explained this approach in the introduction to Part II, further explanation is called for in view of prevalent beliefs about the scientific method. We deeply believe that sociology and the social sciences can become the kinds of sciences that yield the rapid cumulative development of substantive knowledge as well as a platform on which increasingly effective social technologies can be built. This requires that platform to be constructed through using the full range of knowledge within these disciplines, and this in turn requires a language of abstract concepts that are general enough to encompass all that knowledge. Our use of boldface is meant to serve as a reminder of this argument. It is also meant as a reminder of the importance of seeing all of the concepts in Figure 6-1 as being invoked when any one of them is invoked. Further, it is meant as a reminder that we can come to see the ordinary con-

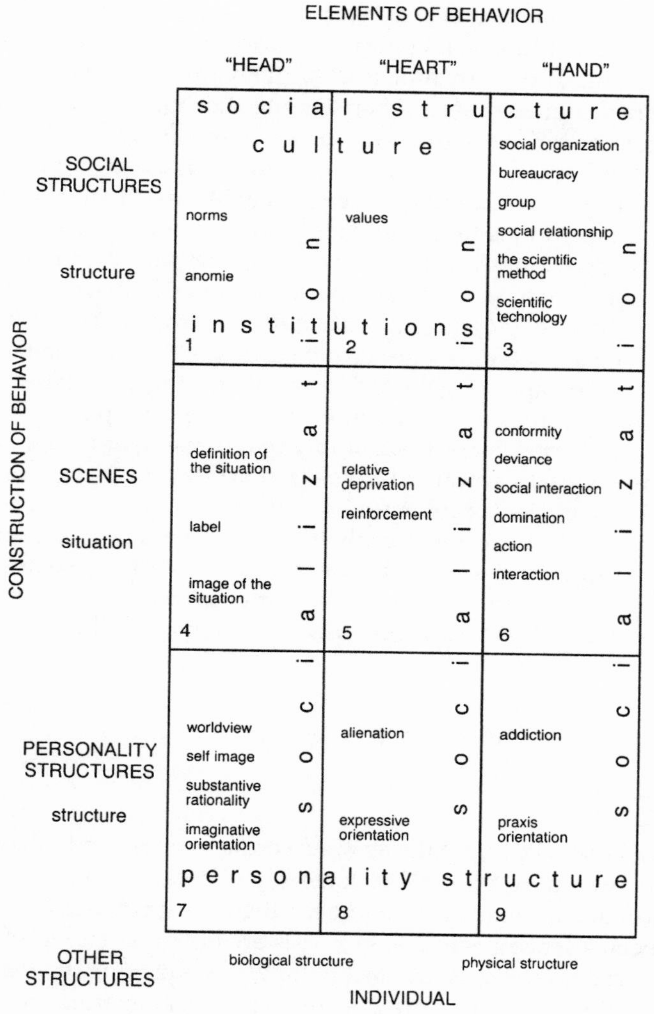

ELEMENTS OF BEHAVIOR

	"HEAD"	"HEART"	"HAND"
SOCIAL STRUCTURES structure	s o c i a l s t r u c t u r e c u l t u r e norms anomie i n s t i t u t i o n s 1	values 2	social organization bureaucracy group social relationship the scientific method scientific technology 3
SCENES situation	definition of the situation label image of the situation 4	relative deprivation reinforcement 5	conformity deviance social interaction domination action interaction 6
PERSONALITY STRUCTURES structure	worldview self image substantive rationality imaginative orientation p e r s o n a l i t y s t r u c t u r e 7	alienation expressive orientation 8	addiction praxis orientation 9
OTHER STRUCTURES	biological structure	physical structure	

CONSTRUCTION OF BEHAVIOR

INDIVIDUAL

Figure 6-1. Elements and construction of behavior: a range of concepts.

cepts we use in everyday life in this way: they are incomplete in describing the phenomena they point toward to the extent that all nine categories of Figures 5-1 or 6-1 are not invoked. The approach here is much like that of Alfred Korzybski (1933) in his emphasis on "consciousness of abstraction": language's concepts, *all* of which are abstract to a degree, are no more than very limited maps of highly complex territories.

In the first section of this chapter, Strengthening Linguistic Tools, we introduce eleven additional concepts to add to the twenty-eight previously employed. Although almost all of them are familiar to the sociologist, they are designed to strengthen the interdisciplinary potential of our entire system of concepts. In addition, we return to our feedback loops, as depicted in Figures 1-4, 1-5 and 4-1. However, by seeing them in the context of our usage of language in everyday life, we are able to point toward very simple loops that help us to penetrate the complexity of ordinary phenomena. Finally, we look to several suggestive metaphors from science fiction to help us understand the nature and impact of language. Our focus is on how language can be changed into a more powerful tool for solving social problems. The second section of the chapter, Back to the Future, centers on applying the full range of our linguistic tools to earlier materials. It has three parts, corresponding to the three parts of this manuscript. For a sharper focus than simply that of using our linguistic tools on the range of earlier materials, we concentrate on what those linguistic tools have revealed about emotional life in modern society. For example, it appears that a central problem is our tendency to repress our emotions, largely in response to the enormous gap between what we want and are able to get. Finally, a very brief section conveys some concluding thoughts about the arguments presented by the book as a whole.

STRENGTHENING LINGUISTIC TOOLS

Abstract Concepts

We have been able to use the abstract concepts depicted in Figure 1-3 in a variety of examples, and—when taken together—they appear to provide a much broader approach to phenomena than we are able to find throughout the social sciences. Yet those concepts are not good enough for highly effective and cumulative social technology. For one thing, they are generally oriented in the direction of a bureaucratic as distinct from an interactive worldview, as illustrated by the concepts of anomie, social stratification, bureaucracy, label, relative deprivation, conformity, alienation, and addiction. This is understandable, since those concepts reflect what *is* more than what *might be*. Yet if we wish to change what is, then we would do well to develop more concepts pointing toward what might be. For another thing, given the complexity of phenomena, what we require is a truly interdisciplinary approach. Granted that the concepts within Figure 1-3 reach out very widely, they are nevertheless quite unbalanced in their emphasis on sociology and the social sciences. It is that imbalance that we must address to an extent. For example, up to this point we have said little about the humanities, yet it is within the humanities more than the social sciences that language is emphasized. If language is our most powerful problem-solving tool, then we cannot afford to ignore the knowledge that has already been developed as to the nature and impact of language.

Let us begin by introducing a number of concepts pointing toward an interactive worldview, and here the metaphors of the "head," the "heart," and the "hand" taken from Figure 5-1 can help. Corresponding to those metaphors, we might introduce the concepts of "imaginative orientation," "expressive orientation," and "praxis orientation" as structures the individual might use to move toward an interactive worldview. We define an **imaginative orientation** as "the individual's openness to learning," taking advantage of the capacities that language has given us humans. We note here the relationship between this concept and Mills's idea of the "sociological imagination," where the individual has learned to move beyond the boundaries of narrow specialization. As for an **expressive orientation**, let us define that concept as "the individual's commitment to awareness and expression of emotions." Granting the importance of alienation throughout modern society, the concept of expressive orientation suggests a way out of alienation, picking up the thread of our analysis of emotions in Chapter 4. As for the "hand," we can define a **praxis orientation**—an approach popularized by Marx—as "the individual's commitment to interaction that shapes self and world." This interactive focus on change contrasts with "addiction," where individuality is subordinated to externally oriented and repetitious behavior.

Yet another structure within the individual can be derived from Weber's contrast between "formal rationality," based on accurate predictions of factors relating to a narrow range of economic values, and "**substantive rationality**," which we may define as "the individual's orientation to the full range of cultural values" (Weber 1964:35–36, 184–85; see also Kincaid 2000). When Weber proceeded to laud the advantages of modern bureaucracy in their achievement of rationality, he apparently gave short shrift to the importance of substantive rationality in favor of formal rationality. Such a broader orientation is illustrated by Constas (1958) and Udy (1959), who distinguished between the stratified aspects of bureaucratic organization and the rational or scientific aspects. Constas and Udy provide an alternative view of bureaucracy that does not, as in the case of Weber, center on predictions based on few cultural values. Rather, they suggest the possibility of increased rationality associated with greater interaction up and down an organizational hierarchy as well as across specialized fields. Thus, substantive rationality gives us a way of challenging the supposed efficiency of bureaucracy and the bureaucratic worldview. Also, it links individual structures with the range of cultural values. Further, it ties in with two concepts we shall now proceed to define: "the scientific method" and "scientific technology."

We define the **scientific method** as "a procedure for achieving deepening understanding of problems that builds on prior knowledge and is gained through patterns of social interaction." Correspondingly, we define **scientific technology** as "a procedure for solving problems that builds on prior knowledge and is gained through patterns of social interaction." We should note here that both the scientific method and scientific technology follow the metaphor attributed to Newton of standing on the shoulders of giants, that is, they involve "building on prior knowledge." Here, the development of literacy and the invention of the printing press provided a basis for taking prior knowledge into account. Further, they both involve "patterns of social interaction" as procedures for gaining such prior knowledge. Here, the university and the professional association have provided key contexts for such patterns of social interaction. This definition of the scientific method is broad enough to encompass the five aspects of that method outlined in Chapters 1 and 2. Those aspects focus on specific procedures for building on prior knowledge. In this definition we add the factor of social interaction, taking into account the importance of the group or community as well. Yet this definition also takes into account postmodernist critiques of the supposed infallibility of the scientific method: there is no claim here that the scientific method necessarily achieves truth or moves us inevitably closer and closer to truth.

These new concepts help us to see more clearly the road that might be taken to the development of an interactive worldview and cultural paradigm, pointing us away from such phenomena as anomie, alienation, and addiction. However, we also require concepts that strengthen what we have learned from disciplines outside the social sciences. Edward O. Wilson, building on the very broad orientation of Sir Francis Bacon, sketches the direction that he believes is essential for finding a way out of the failures of the Enlightenment:

> True reform will aim at the consilience of science with the social sciences and the humanities in scholarship and teaching. Every college student should be able to answer this question: What is the relation between science and the humanities, and how is it important for human welfare? . . . Most of the issues that vex humanity daily—ethnic conflict, arms escalation, overpopulation, abortion, environmental destruction, and endemic poverty, to cite several of the most persistent—can be solved only by integrating knowledge from the natural sciences with that from the social sciences and the humanities. Only fluency across the boundaries will provide a clear view of the world as it really is, not as it appears through the lens of ideology and religious dogma, or as a myopic response solely to immediate need. . . . A balanced perspective

cannot be acquired by studying disciplines in pieces; the consilience among them must be pursued. Such unification will be difficult to achieve. But I think it is inevitable. (1998:62)

Granting that we already have the concepts of biological structure and physical structure, we need reinforcements in beginning to open up more fully to the biological and physical sciences. The concept of **structure**—"a persisting system of elements"—helps us to link physical, biological, social, and personality structures, given its inclusive nature, getting away from a sharp dichotomy between the social sciences and other sciences. And the concept of **situation**—"any phenomenon located in time and space"—does the same general thing. And when situation and structure are combined, we open up to an understanding of change and not just social and personality change. For example, differential equations, which are so fundamental to the physical sciences, depend on that linkage. Two other concepts help us to pull in the literature of psychology. We define **action** simply as "individual behavior," and **interaction** as "individual action that yields environmental response." Our present concept of social interaction becomes, then, one type of action. **"Action"** also refers to behavior of the human being, but not necessarily social interaction, opening up to any thoughts, feelings, or physical behavior of the individual. **"Interaction,"** similarly, does not require social interaction, but merely any response from the environment derived from an action, such as a computer's feedback from an input. Again, we take away from emphasizing an exclusive sociological focus on social interaction.

As for linkage with the humanities, figurative language—communicating in nonliteral ways—such as by metaphor, simile, personification, and metonymy, is central to poetry and literature. These are ways of saying something more vividly and forcefully, as illustrated by Nietzsche's metaphor, "God is dead," and they are used in everyday speech as well. The impact on us of figurative language conveying images, in common with descriptive language that paints pictures, has a basis in the importance for us of perception in everyday life. In common with other organisms, our perception is central for our survival. Yet by contrast with other organisms, language gives us a basis for greatly extending our powers of perception. We have already begun to recognize the importance of images in our concepts of "worldview," "self-image," and "imaginative orientation." However, all of these refer to structures within the individual and not to the momentary scene. Let us then introduce an additional concept bearing on that scene. **Image of the situation** parallels the sociological concept "definition of the situation," which was defined as the individual's understanding of the momentary scene. Analogously, we define image of the situation as "the individual's view of the momentary scene." In this way we tap into a powerful device of poetry and literature as well as ordinary language, opening up to our capacity for perception.

Figure 6-1 presents the range of concepts emphasized in this book using the same format as Figure 5-1. In this way we can see those concepts as illustrative of the nine categories of 5-1 and we can locate other concepts as well within those categories. The basic idea behind Figures 6-1 and 5-1 is much the same: that we should learn to use more of those categories than we generally do, since phenomena are more complex and dynamic than we generally assume. By increasing the number of categories we employ we open up to more of the potential that language has to offer us. For example, by including aspects of both structures—located in the first and third rows as well as below the nine categories—and momentary situations, we are able to learn more about social change. If an occurrence within a scene is repeated over and over this can produce a structure, and correspondingly a structure influences what occurs within any given scene. Beyond helping us to understand social change, multiple categories can help us to achieve social change. For example, an imaginative orientation can help us to develop an alternative cultural paradigm, an expressive orientation can aid in our becoming motivated to move in that direction, and a praxis orientation can help us to actually initiate such movement.

Causal-Loop Diagrams

Given our socialization to a bureaucratic worldview with its narrow approach to language, how can we learn to open up to using multiple categories in our everyday thoughts and speech? Granting the importance of dipping into both structures and the situation in order to develop a more dynamic perspective, how can we proceed to learn to dip into both? It is exactly here that the causal-loop diagram can help us (see especially Roberts et al. 1983:1–86), as exemplified by Figures 1-4, 1-5, and 4-1. Those figures give us images of links among concepts, links that take us across multiple categories within Figures 5-1 and 6-1. Such images are more literal than the figurative images within linguistic metaphors and descriptive language, and they affect our perception more directly. As a result they appear to have great potential for helping us learn to think in more complex and dynamic ways. In the preceding materials we have used causal feedback loops to present fairly complicated arguments relating to the sociological literature in a systematic fashion. Our focus in this section is not on such complicated arguments but rather on relatively simple causal feedback loops that can help us carry our thinking just a bit further in the direction of complexity and dynamism. Instead of diagrams with some ten to twenty concepts, we will generally limit ourselves to two or three. Yet even such simplicity represents a more complex and dynamic way of thinking than we currently use.

In addition to this simplicity there is also the question of the level of abstraction we employ in the concepts we use. Our emphasis in the preceding chapters, appropriate to our focus on integrating ideas from the literature of sociology, has been on very abstract concepts, as illustrated by all of the concepts in Figure 6-1. Such abstraction is crucial to our ability to pull together the literature of sociology and the social sciences, and it appears to be basic to our web approach to the scientific method. Yet this is certainly not the way any of us think, and the Ph.D. degree makes little difference here, given the socialization we all experience in everyday life. What we require *in addition to* such abstract concepts are illustrations that take us far down language's ladder of abstraction to our everyday usages. We must learn how to link less abstract concepts like "fear," "shame," "confidence," and "love" to those abstract concepts in order to make better use of language's potential. Thus, our simple causal-loop diagrams should not be restricted to the concepts listed in Figure 6-1. Of course, this is by no means a solution to the problem of how to think in more complex and dynamic ways in everyday life. But it does appear to point us in this direction. It takes us further toward using the power of an interactive scientific method in our everyday thoughts, granting the long road ahead in resocializing ourselves to think in more interactive ways.

Causal-loop diagrams can help us in yet another way. A glance at Figure 6-1 with its emphasis on moving toward using all nine categories in the analysis of any phenomenon illustrates enormous complexity. And that complexity is greatly increased when we include concepts at lower levels of abstraction and, further, when we introduce a reflexive orientation to our analysis. It is little wonder that the traditional bureaucratic approach to the scientific method within the social sciences has thrown up its hands in the face of this problem and escaped into the kind of highly specialized work that takes little responsibility for knowledge within social science as a whole. Causal-loop diagrams do not suddenly solve this incredibly difficult problem, but they do offer us a direction for learning to think in ever more complex ways as we proceed to introduce more concepts into our diagrams. And this direction leads us toward procedures for computer simulation that can link our concepts in highly systematic ways and also assess their utility for explaining concrete data, all in accordance with our web approach to the scientific method. Yet we need not repeat the mistake made by so much of positivistic methodology by centering on formal rationality to the exclusion of **substantive rationality**. Following Gandhi's approach to *satyagraha*, causal-loop diagrams can be an end as well as a means leading in the long run toward computer simulation.

To illustrate this simplified approach to causal-loop diagrams, let us return to Chapter 4's analysis of Gandhi's salt *satyagraha* in early May 1930, involving a march to Dharsana to demand possession of the large salt works located there. Gandhi communicated his plan to the British viceroy on April 17 and was subsequently arrested on May 5, but leading Congress party officials followed by volunteers proceeded without him. Let us repeat the passage describing the police reaction to the march:

> An American journalist, Negley Farson, recorded an incident in which a Sikh, blood-soaked from the assault of a police sergeant, fell under a heavy blow. Congress first-aid volunteers rushed up to rub his face with ice. . . . "He gave us a bloody grin and stood up to receive some more." . . . The police sergeant was "so sweaty from his exertions that his Sam Browne had stained his white tunic. I watched him with my heart in my mouth. He drew back his arm for a final swing—and then he dropped his hands down by his side. 'It's no use,' he said, turning to me with half an apologetic grin. 'You can't hit a bugger when he stands up to you like that!' He gave the Sikh a mock salute and walked off." (Bondurant 1965:96)

We can proceed to contrast here the behavior of the British police sergeant with that of the Sikh volunteer. And in the process we can employ both technical concepts from Figure 6-1 and the everyday concepts we use in ordinary thought and speech.

Figure 6-2 presents this contrast. We know very little about just what the British police sergeant was thinking and feeling as he continued to beat the volunteers, yet we may suggest a series of hypotheses on the basis of the general knowledge we do have. We may view him as initially deriving positive **reinforcement** from **conformity** to the orders given to him by his superiors. These **situational** concepts enable him, temporarily, to blot out from his mind his violation of deep **cultural values** that have a basis in the Western religious **institution,** such as "equality," "freedom," and "individual personality" or the ultimate worth of the individual. His **action** of **domination** is also supported by fundamental **social structures:** patterns of **social organization,** such as **social stratification** and **bureaucracy,** and also his **social relationships** within his military **group.** Other **structures** are also involved, namely, **personality structures,** such as the sergeant's **worldview** and **self-image.** We can also make this same argument in nontechnical language. For example, we can see the sergeant's behavior as illustrating obedience to authority, and we can see him deriving personal satisfaction from his following the rules.

These simple positive loops yield acceleration of behavior in the initial direction, where conformity yields reinforcement, reinforcement produces more conformity, and so on. But they appear to be short-lived, for deep contradictions soon emerge, expressed in the form of **alienation.** And this results in converting the positive loop into a negative loop where there is

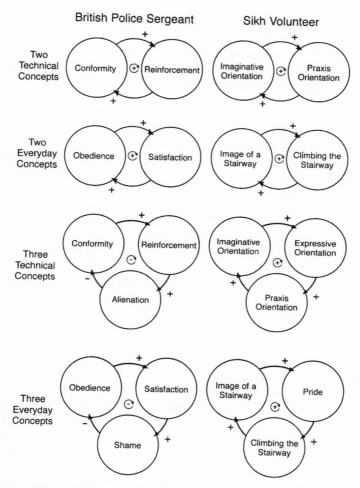

Figure 6-2. *Satyagraha* in Dharsana. Causal-loop diagrams.

a reversal of behavior away from its initial direction. The sergeant becomes more and more aware that he is violating basic **cultural values,** and those violations yield negative **reinforcements** along with alienation when he continues attempting to obey orders: "He drew back his arm for a final swing—and then he dropped his hands down by his side. 'It's no use,' he said, turning to me with half an apologetic grin. 'You can't hit a bugger when he stands up to you like that!' He gave the Sikh a mock salute and walked off." The mock salute is a gesture of respect for cultural values they both shared, such as "freedom" and "individual personality." The apologetic grin appears to indicate the sergeant's feelings of alienation for what he had been doing. Going around the loop with everyday concepts, the concept of shame may be applied to those feelings, and such feelings of shame take away from the satisfaction he has been feeling as a result of his obedience to the rules. And this in turn leads to his walking away from the job his superiors expect him to do.

Let us now shift our focus to the Sikh volunteer. How can we possibly explain his near-suicidal behavior as he continued to stand up to receive more blows, which could cripple him for life or kill him? Again, with no direct knowledge of his thoughts and feelings, we can only hypothesize, yet those hypotheses stem from our web of sociological knowledge about the nature of modern society. Just as Einstein had to have an alternative paradigm in order to question the existing Newtonian one, so does the Sikh volunteer employ his **imaginative orientation** to **define the situation** he is in much differently from the way it might appear to most observers. Whereas they might see him as risking enormous negative **reinforcements** in an irrational way, his **image of the situation** appears to be that of gaining positive **reinforcements** by sacrificing his well-being and perhaps his life, following his **praxis orientation,** to oppose further British rule and help gain independence for India. Moving to the second row of Figure 6-2, he might imagine himself climbing a stairway leading to the achievement of that independence. He too illustrates **conformity** to the expectations of a **group,** and in this way he too follows patterns of **social stratification** and **bureaucracy,** but his is not the military group of the police sergeant but the group of volunteers led by the officials of the Indian Congress party as well as by Mohandas Gandhi in absentia.

If we move to the third and fourth rows of Figure 6-2, we have a new concept corresponding metaphorically to the "heart" and thus completing the head-heart-hand trilogy that encompasses a wide range of human behavior. In order to do what he did the Sikh volunteer had to be very highly motivated, and we might assume that he had developed an **expressive orientation** enabling him to bring forward that motivation in his confrontation with the British police sergeant. If we follow the argument in earlier chapters, then—like the Springdalers—we moderns have learned to suppress our feelings to a great degree, given the large gap between our aspirations or **cultural values** and our ability to fulfill them, illustrated by our widespread **alienation.** Yet an expressive orientation points us toward the possibility of fulfilling those cultural values and moving away from such alienation. Such an orientation helps to give the Sikh volunteer the enormous sense of problem that he requires to risk his life for the cause of independence. And apparently he is able to envision the effectiveness of the links between his **imaginative orientation, expressive orientation,** and **praxis orientation** in helping to generate forces that will ultimately achieve

Indian independence. We can see the sergeant's giving up in the face of the volunteer's self-sacrifice as a metaphor for Britain's ultimate relinquishing of its control of India.

These simple causal-loop diagrams only take us a limited distance toward understanding the complexity of what is happening. By introducing other concepts in this discussion we take into account more of that complexity. Yet from a pragmatic perspective in an effort to learn to think in more complex ways, what is crucial is how those simple diagrams compare with our ways of thinking in everyday life. For example, it appears that we generally do not take into account both long-term **structures** as well as what is occurring in the immediate **situation.** And as a result we have great difficulty when it comes to thinking in dynamic ways, especially over very long periods of time such as the change from preindustrial to modern society or from oral culture to literate culture. The feedback loop idea centers on change, for we continue to go around the loop without any end. Further, it can focus on change within the situation at hand, as in the case of the conformity-reinforcement or obedience-satisfaction loops for the British sergeant. Also, it can focus on long-term structures, as in the case of the volunteer's imaginative orientation–praxis orientation. And it can also encompass both structures and situations, as in the case of the British sergeant's conformity-reinforcement-alienation loop. As a result, by learning to analyze phenomena with such very simple loops we are able to learn to think in more complex and dynamic ways, granting the simplicity of those causal-loop diagrams.

Allegories for Understanding Language

A figure of speech is a way of saying something in a nonliteral way, and a metaphor is a way of comparing things that are essentially unlike. When Nietzsche says that God is dead he does not mean to claim, literally, that some individual who is named God has died. Rather, he is comparing ideas about the general nature of God with ideas about the general nature of death. A nonliteral reading of his statement can yield several meanings other than the literal one that God is dead. For example: belief in God is no longer useful or is even harmful; people no longer believe in God; belief in any elite figure or hierarchy is no longer appropriate; and our cultural paradigm based in part on belief in God no longer works. Actually, Nietzsche's statement is a subtype of metaphor, namely, a personification. It is, just like any metaphor, a comparison between two things, but one of them is not human. In this way the attributes of a human being—like dying—are given to some nonhuman object or idea. If Nietzsche had said that God is *like* a dead person he would be using a simile, which is yet another type of metaphor. The comparison Nietzsche is invoking, like any type or example of metaphor, enables us to envision someone with the appearance of God lying very still in a coffin. Metaphor enables us to see the world in a certain way, one that is different from our ordinary mode of perception. As a result it can affect us profoundly with a very few words, perhaps the basis for its extensive use in poetry.

By looking to metaphors in this subsection, we are continuing our effort to develop powerful visual tools that can help us to understand language more fully and also understand how language can be used more effectively for social technology. We might recall here our introduction of the concepts "image of the situation" and "imaginative orientation" in the first subsection. Precious few of the thirty-nine concepts in Figure 6-1 deal with images: in addition to those concepts there is only "self-image" and "worldview." In our second subsection we introduced another visual device: the causal-loop diagram. And the figures we used in earlier chapters have enabled us to gain clarity with respect to very abstract ideas. Our focus in this subsection will be on two allegories, which we may view as systems of extended metaphors conveyed by stories that have a second meaning beneath the surface one. George Orwell's *Nineteen Eighty-Four* (1949) does not center on language, but a significant portion of this science-fiction novel has to do with a comparison of "Newspeak," the language of the Party, with "Oldspeak." Orwell's focus on Big Brother is not just a metaphor for the horrors of Stalinism but also for the horrors of modern bureaucratic society. Jack Vance's *The Languages of Pao* (1958) may be taken to be an allegory for historical change from oral society to literate industrial society and then to a society based on an interactive worldview. And it is change in language that is the fundamental basis for change in society.

Nineteen Eighty-Four

Orwell's novel is an effort to convey a self-defeating prophecy: by revealing the horrors embodied in the directions being taken by modern society he hoped to sound a warning before it is too late to reverse the process. Winston Smith is the protagonist in the future society of Oceania, a dictatorship far more totalitarian than any that has ever existed. Two-way telescreens and informers everywhere detect the least lack of enthusiasm of the individual, and unspeakable torture converts those without enthusiasm into automatons who worship Big Brother. Changing the existing language of Oldspeak to Newspeak was a key device used by Big Brother— who resembled Stalin—to achieve totalitarian ends. Our own focus here is by no means on a pessimistic view of our future. Rather, Orwell's insights into how language can be shaped so as to destroy the individual's humanity can also yield insight into the reverse process: how language can be changed so as to enhance our humanity and help us confront the basic problems of modern society. Orwell's novel emphasizes the evils of dictatorship within a highly stratified and bureaucratic society, yet most of what he says can also apply to modern society as a whole with its emphasis on a bureaucratic worldview, bureaucracy, social stratification, and conformity to cultural values and norms. Orwell is not discussing our inevitable fate but rather a fate that awaits us all unless we change it.

Winston's colleague in the Ministry of Truth, Syme, is one of the team compiling the definitive edition of the Newspeak dictionary:

> You think, I dare say, that our chief job is inventing new words. But not a bit of it! We're destroying words—scores of them, hundreds of them every day. . . . Do you know that Newspeak is the only language in the world whose vocabulary gets smaller every year? . . . Don't you see that the whole aim of Newspeak is to narrow the range of thought? In the end we shall make thoughtcrime literally impossible, because there will be no words in which to express it. . . . Every year fewer and fewer words, and the range of consciousness always a little smaller. . . . By 2050—earlier, probably—all real knowledge of Oldspeak will have disappeared. The whole literature of the past will have been destroyed. Chaucer, Shakespeare, Milton, Byron—they'll exist only in Newspeak versions, not merely changed into something different, but actually changed into something contradictory of what they used to be. Even the literature of the Party will change. Even the slogans will change. How could you have a slogan like "freedom is slavery" when the concept of freedom has been abolished? The whole climate of thought will be different. In fact there will *be* no thought, as we understand it now. Orthodoxy means not thinking—not needing to think. Orthodoxy is unconsciousness. (1949: 45–47)

Orwell emphasizes the dictator's role in creating a dehumanized society, yet we need not adopt that focus. Instead, we can look to the forces in society that limit or expand our consciousness.

Our overall emphasis on an accelerating gap between aspirations and fulfillment in modern society suggests—failing some direction that narrows that gap—the repression of any consciousness of that gap. Vidich and Bensman analyzed Springdalers who illustrated that gap in their own lives and achieved such repression through techniques of particularization analogous to the narrow specialization of social scientists. And those Springdalers also limited their consciousness through the falsification of memory, just as social scientists in general no longer emphasize the Enlightenment dream of penetrating the nature of modern society's fundamental problems. These procedures appear to work in the same direction as Big Brother's creation of Newspeak, namely, to "narrow the range of thought," limit "the range of consciousness," and move toward "unconsciousness." To achieve this within the social sciences it is not necessary to eliminate words. Instead, we need only emphasize concepts at a low level of abstraction, by contrast with the web approach to the scientific method. Such an emphasis makes it very difficult to link the knowledge in one subfield with the knowledge in other subfields, let alone link knowledge among disciplines. The cliché that we are learning more and more about less and less applies very well to this approach. We appear to be creating Newspeak without the benefit of Big Brother.

Does our web approach to the scientific method point us away from techniques of particularization, the falsification of memory, and the narrowing of our range of thought and consciousness? Does it not only point us away from Newspeak but also toward changing our Oldspeak so that we learn to make use of ever more of the knowledge of the social sciences within our everyday thoughts? Apparently so. Just as using concepts at a low level of abstraction works to limit the range of our thought, so do abstract concepts open up that range of thought and consciousness. And in the same way, when those abstract concepts are linked together systematically, when we are also encouraged to move down the ladder of abstraction, and when we are committed to adopting a reflexive orientation, we continue to expand our range of thought and consciousness. Further, that expansion continues when we are encouraged, as in the case of the web approach, to take up the large problems of society and self. But to accomplish all of this we require an alternative paradigm, a direction for resolving the problems that would appear were consciousness to expand rather than contract. Although the Springdalers—along with the rest of us—were able to obtain at least "some degree of satisfaction, recognition and achievement" through their contraction of consciousness, a great deal more "satisfaction, recognition and achievement" appears to await us if we can move toward an alternative paradigm and learn to expand our consciousness.

The Languages of Pao

Jack Vance's *The Languages of Pao* (1958) gives us a view of the power of language for changing the world. If we rewrote our history books so as to emphasize the role of language rather than wars, religious and political leaders, cultural events, exploration, the formation of nations, and the development of inventions, then we might distinguish two major periods: preliterate and literate. The first would encompass almost all of human history, would include hunting-and-gathering as well as the invention of agriculture, and would involve little *fundamental* change in the individual's ability to shape the environment. The second would, within a few millennia, yield the scientific and industrial revolutions of the past four centuries and succeed in threatening the future of the human race and perhaps of all life as well. *The Languages of Pao* may be viewed as an allegory showing how language has taken us from the first to the second period. And it is also an allegory suggesting how language might be employed so as to take us into a third period where we are no longer threatened by the so-called advances of technologies based on the physical and biological sciences. Vance's book, then, can give us metaphors that can help us to understand not only the nature of language but also the possibility of using language— just as we are attempting with our web approach to the scientific method— as a social technology for changing the world.

Vance's novel is based on the linguistic relativity hypothesis of Benjamin Lee Whorf and Edward Sapir: that language causes people to understand the world in a certain way (Whorf 1963). He carries forward the implications of this thesis to include not just understanding or the "head" but also the "heart" and the "hand": the language we use also shapes the way we feel and act. Vance sets his tale mainly on the planet Pao, a world of fifteen billion inhabitants originally colonized from Earth and existing in the far future. The neighboring planets of Breakness, Mercantil, and Batmarsh were the locations of societies differing markedly from Paonese society. Breakness was a land of male scientists, importing females from neighboring planets solely for the purpose of procreating males and shipping the women back along with their female children as well as their male children who were unable to gain entrance as students in the Breakness Institute. Mercantil was a merchant planet that produced a wide range of goods and was home to the key traders in the area, with their fleet of ships. Batmarsh was the home of warrior clans who competed with one another and raided neighboring planets seeking tribute. By contrast, the Paonese had not developed in such specialized ways. They expected little from life and emphasized caste or status and tradition. They had no competitive sports, typically farmed a small acreage, and gave total allegiance to the Panarch, the hereditary ruler who reached out throughout Pao with a vast civil service.

Beran Panasper, the young son of the Panarch and heir to the throne, witnesses his father's assassination by his uncle, Bustamonte, who blames two emissaries from Mercantil for the deed and promptly executes them. Lord Palafox, Wizard of Breakness, witnesses what has occurred and escapes to the planet Breakness with Beran before Bustamonte is able to act against them. Bustamonte proclaims Beran's death and assumes the throne of Pao while Beran becomes a student at the Breakness Institute under Lord Palafox's patronage. Meanwhile, Pao continues to be invaded by the Brumbos of Batmarsh, who require ever higher amounts of tribute. The Paonese, despite their population of fifteen billion, are a passive people who are unable to mount a defense, and Bustamonte travels to Breakness to seek Palafox's help. Palafox, Dominie of Comparative Culture, comes up with a general plan:

> We must alter the mental framework of the Paonese people—a certain proportion of them, at least—which is most easily achieved by altering the language. (Vance 1958:57)

Paonese on three continents would be relocated to make way for children who would learn a new language. Valiant, Technicant, and Cogitant would, over time, yield warriors, industrialists, and scientists superior to those on Batmarsh, Mercantil, and Breakness.

Palafox explains the rationale behind his plan:

> Paonese is a passive, dispassionate language. It presents the world in two dimensions, without tension or contrast. A people speaking Paonese, theoretically, ought to be docile, passive, without strong personality development—in fact, exactly as the Paonese people are. The new language [Valiant] will be based on the contrast and comparison of strength, with a grammar simple and direct. To illustrate, consider the sentence, "The farmer chops down a tree" [literally, in Paonese, "Farmer *in state of exertion;* axe *agency;* tree *in state of subjection to attack*]." In the new language the sentence becomes: . . . "The farmer vanquishes the tree, using the weapon-instrument of the axe". . . . The syllabary will be rich in effort-producing gutturals and hard vowels. A number of key ideas will be synonymous; such as *pleasure* and *overcoming a resistance—relaxation* and *shame—out-worlder* and *rival.* Even the clans of Batmarsh will seem mild compared to the future Paonese military.

Another area might be set aside for the inculcation of another language [Technicant]. . . . In this instance, the grammar will be extravagantly complicated but altogether consistent and logical. The vocables would be discrete but joined and fitted by elaborate rules of concordance. What is the result? When a group of people, impregnated with these stimuli, are presented with supplies and facilities, industrial development is inevitable. . . . To the military segment, a "successful man" will be synonymous with "winner of a fierce contest." To the industrialists it will mean "efficient fabricator." (ibid.:58–59)

As for Cogitant, the language of the scientist, it was similar to the language of Breakness, emphasizing the "head" with almost nothing in the area of the "heart." Beran relates his initial experiences of learning the language on the planet Breakness:

> The language included no negativity; instead there were numerous polarities such as "go" and "stay." There was no passive voice—every verbal idea

was self-contained: "to strike," "to receive-impact." The language was rich in words for intellectual manipulation, but almost totally deficient in descriptives of various emotional states. Even if a Breakness dominie chose to . . . reveal his mood, he would be forced to the use of clumsy circumlocution.

Such common Paonese concepts as "anger," "joy," "love," "grief," were absent from the Breakness vocabulary. On the other hand, there were words to define a hundred different types of ratiocination, subtleties unknown to the Paonese . . .

On Pao there was small distinction between the sexes; both wore similar garments and enjoyed identical privileges. Here the differences were emphasized. Men wore dark suits of close-fitting fabric. . . . Those whom Beran had glimpsed wore flouncing skirts of gay colors—the only color to be seen on Breakness . . . all were young and handsome. . . . As he stood, a group of boys several years older than himself came up the road from the Institute; they swerved up the hill, marching in a solemn line. . . . Curious! thought Beran. How unsmiling and silent they seemed. Paonese lads would have been skipping and skylarking. (ibid.:42–43)

Beran joins a group of Paonese on Breakness who are learning all three languages and who, as the only interpreters able to communicate with everyone, will assist the Paonese civil service. During their studies they proceed on their own in a jocular vein to invent a patchwork tongue made up of scraps of Valiant, Technicant, Cogitant, and Paonese, which they christen "Pastiche." Returning to Pao the interpreters assist in the education of the enclaves of Paonese children. And after only a single generation Palafox's plans bear fruit. Paonese warriors who had learned Valiant and had developed their own warrior traditions outfought the clansmen of Batmarsh. Paonese entrepreneurs and traders, students of Technicant, proved to be superior to their counterparts on Mercantil. And Paonese intellectuals compared favorably with the so-called Wizards of Breakness. Beran replaces Bustamonte as Panarch and successfully opposes Palafox, both of whom were attempting to institute an authoritarian regime on the planet. Yet he finally realizes that the three separate enclaves have developed loyalty only to their own groups and none to the Paonese as a whole. For example, the Valiants oppose Beran's orders and threaten his life. Beran decides to disperse the Valiants, Technicants, and Cogitants into small groups who will educate and train their fellow Paonese in their own knowledge and skills. And the Paonese will learn a single language enabling everyone to combine all of these achievements: Pastiche.

Viewing Vance's tale as an allegory of the change from preliterate and preindustrial to modern society yields several insights. Although his linguistic focus oversimplifies the many factors involved in that vast change, nevertheless it points up an idea to which we give too little attention: the centrality of language for cultural change. We might view Paonese language and culture as analogous to preliterate society, with its emphasis on tradition and the passivity of the population at large. A lack of emphasis on abstract thought (Ong 1982:especially 49–56; see also Luria 1976) along with a corresponding focus on the momentary scene is very far from the mentality of the scientists and the physical and biological technologists who have come to the fore from the seventeenth century up to the present. Yet that focus on the momentary scene also suggests a balance among "head," "heart," and "hand," for immersion in the scene requires attention to all three. With the development of literacy along with the widespread availability of written materials, it was possible for the individual to move far away from the scene in which he or she was located. Here, we can use Vance's description of the Cogitants and Technicants as metaphors for what has in fact occurred within our scientific and industrial revolutions. For example, we might recall the Breakness students—similar to the Cogitants—and their lack of laughter and somber colors, given a language without words for emotions.

The Languages of Pao also gives us an allegory for a change from modern society with its failures of communication—as illustrated by the forty sections of the American Sociological Association—to a far more communicative society. Vance's solution to this problem is that all individuals learn one language—Pastiche—which is sufficiently broad and flexible so as to embody the ideas and ideals of the Valiants, Technicants, Cogitants, and Paonese. Pastiche contains elements from all of the other languages and somehow manages to make them all relevant. Given such inclusiveness, it cannot succumb to the one-sidedness of Valiant, Technicant, and Cogitant. And neither can it succumb to the passivity of Paonese. We might see it as suggesting a return to the balance among "head," "heart," and "hand" to be found within preliterate society, yet also somehow including the achievements of modern society. It appears that Pastiche is similar to the linguistic orientation within a web approach to the scientific method. Our emphasis on very abstract concepts is not for the purpose of moving away from the momentary scene, but rather enables us to immerse ourselves more fully within that scene, taking into account the ideas, feelings, and actions of both speaker and audience. And our emphasis on a systematic web of concepts can enable us to pull together what Pastiche achieves only in a patchwork way: the full range of ideas and interests we moderns embody.

BACK TO THE FUTURE

With the aid of our linguistic tools—the new ones along with those developed previously—we are in a position to return to our earlier chapters and to take another look at our arguments linked to a reconstruction of the scientific method. Overall, those tools point precisely in the direction Nietzsche took in his *The Gay Science:*

> *Taking seriously.*—In the great majority, the intellect is a clumsy, gloomy, creaking machine that is difficult to start. They call it "taking the matter *seriously*" when they want to work with this machine and think well. How burdensome they must find good thinking! The lovely human beast always seems to lose its good spirits when it thinks well; it becomes "serious." And "where laughter and gaiety are found, thinking does not amount to anything": that is the prejudice of this serious beast against all "gay science."— Well then, let us prove that this is a prejudice. ([1887] 1974:257)

"Laughter and gaiety" illustrate emotional expression, or attention to the "heart" and not just the "head" and the "hand." If the scientific method is to invoke, reflexively, the deepest possible emotional commitment of the investigator, then it cannot afford to rule out "gaiety" in favor of "taking the matter seriously." Our image of the unemotional scientist, as someone governed by "formal rationality" and thus an excellent calculator, is deficient. **Substantive rationality,** with its opening up to the range of **cultural values,** appears to be a far more appropriate orientation for the scientist.

Such a broad emotional orientation can also help the scientist face up to the deepest problems of modern society:

> *On the aim of science*—What? The aim of science should be to give men as much pleasure and as little displeasure as possible? But what if pleasure and displeasure were so tied together that whoever *wanted* to have as much as possible of one *must* also have as much as possible of the other—that whoever wanted to learn to "jubilate up to the heavens" would also have to be prepared for "depression unto death"? . . . Actually, *science* can promote either goal. So far it may still be better known for its power of depriving man of his joys and making him colder, more like a statue, more stoic. But it might yet be found to be the *great dispenser of pain*. And then its counterforce might be found at the same time: its immense capacity for making new galaxies of joy flare up. (ibid.:85–86)

Here again, Nietzsche points us toward the emotions, not just those of the scientist but also the pleasure and pain to be found in society. Science can address our deepest problems, those which give us the greatest pain, such as the contradiction between the **cultural value** of "equality" and our patterns of **social stratification.** Yet following the pendulum metaphor we have used to characterize a web approach to **the scientific method,** this approach can in turn promise us the greatest pleasure: not only in solving those problems but enabling us to move far beyond them.

In this final section we follow Nietzsche's lead in emphasizing emotional life as we proceed to reexamine Parts I, II, and III. Earlier material included the analysis of emotions, but it did not have this as a focus. Granting that such a focus does not somehow solve the scientific problem of how to develop an **interactive scientific method** and **worldview,** it does seem to be an important pragmatic step in that direction. It is indeed surprising, as we return to earlier parts, just how much of earlier material has centered on the problem of emotional development. The idea of the serious scientist is most appropriate within a bureaucratic approach to the scientific method, where the "two cultures" of the sciences and the humanities are kept far apart. Within that perspective, metaphors and images and language should remain the province of the humanities and not the sciences. And emotions come to be seen as taking away the cold ability to reason, which alone guides us to Truth, much like the way of life and the language on the planet Breakness. But apparently that perspective has failed us in the cumulative development of the social sciences, let alone in the realm of effective social technology. What we appear to require, following Nietzsche, is nothing less than a gay science that will open up to our deepest problems, just as Nietzsche himself suffered great physical pain. And by so doing it promises to enable us to address those problems with effective social technologies.

Part One: The Scientific Method:
Bureaucratic and Interactive Paradigms

Figures 1-1, 1-2, and 1-4 center on perhaps the deepest problems of modern society, with Figures 1-3 and 1-5 focusing on directions for solutions, following Nietzsche's argument about pain and joy as well as our own pendulum metaphor for **the scientific method.** Given the depth of this problem, we Springdalers shunt it outside our awareness with our "techniques of particularization," illustrated by the forty sections of the American Sociological Association. We also escape from this problem through the "falsification of memory," giving up on the Enlightenment dream along with Comte's and Mills's visions of "the promise of sociology." Indeed, it is most threatening for us to visualize a problem that— given our present approach to the scientific method—we have no way to address. And the threat is personal no less than professional, for our **cultural paradigm** and **worldview** are tied to our approach to the scientific method. All of this conspires to make our deepest problems invisible, yielding a far more dangerous situation than that involving the visible enemy we faced during World War II. Yet following Kuhn and his implications, alternative paradigms for sociology and society can help us to make those problems ever more visible. Those paradigms emphasize the use of linguistic tools, such as the abstract concepts and metaphors introduced in Chapter 1. Those tools, in common with visual tools like Figures 1-1 through 1-5 as well as ordinary language, help us to see what we otherwise would fail to see.

We are able to see the emotional dimension within Part I in the stress on addressing the fundamental problems of society, as illustrated within Figures 1-1, 1-2, and 1-4. In Figure 1-1 we are rapidly heading toward disaster as the gap between expectations and fulfillment continues to increase exponentially. It is a growing gap that is also suggested by the link between anomie and modernization, involving basic contradictions between cultural values and patterns of social organization. And these problems at the level of social structure are also reflected in problems at the level of personality structure, such as alienation and addiction. Figure 1-2 emphasizes the linguistic and historical basis for this state of affairs: the greater harnessing of the power of language within the physical and biological sciences by contrast with the social sciences. This has led over the past four centuries to the rapid cumulative development of the former relative to the latter, and also to the development of ever more powerful technologies based on the former. The causal-loop diagram in Figure 1-4 summarizes much of this analysis, pointing up the importance of links between our sociological paradigm and our worldview based on a bureaucratic cultural paradigm. That figure also points up the failure of sociology to fulfill its promise, associated with a corresponding failure to point up an alternative to those paradigms and that worldview. Yet these diagrams offer a direction for that alternative, starting with an assessment of our problems.

We can look to Figure 1-3 as carrying forward the implications of that assessment. Key concepts within that figure center on our emotions: **cultural values, relative deprivation, reinforcement,** and **alienation,** none of which has become central to the literature of sociology. "Reinforcement" is seen by many as tied to psychology. "Cultural values" have never been emphasized as important structures, with sociologists generally following anthropologists in stressing the diversity of cultures and cultural values. "Relative deprivation" has never been linked in any systematic way to **social stratification.** And "alienation" is seen as tied closely to Marxist theory. By contrast, we can use "cultural values" to unearth a wide range of largely hidden forces affecting emotional life throughout modern society. "Relative deprivation" can bring our understanding of social stratification into the momentary scene. "Reinforcement" helps us to make use of a great deal of knowledge developed by psychologists. And alienation, with its focus on the emotional problems of us moderns, remains a fundamental problem of modern society. Figure 1-5 follows Kuhn's approach to the scientific method: to be able to face up to our problems rather than avoid them in Springdaler fashion, we had better develop a direction for solving them.

And to the extent that such a direction becomes a viable one for us, we can deal more and more with emotions like fear, which teach us to avoid our problems.

As we proceed to focus on emotions, let us not skirt over the function of language in general—and these concepts as particular examples—of helping us to bring our emotions to the surface, where they are visible. It is not just an alternative worldview that helps us to face up to our problems but also language. The concepts in Figures 1-3 and 1-5 work together to give us some understanding of our worldview. But those concepts also work individually, just as does ordinary language, in helping us to see what we would miss otherwise. Vision is one thing that language gives us. We also obtain such vision from our schematic diagrams and causal-loop diagrams. It is the invisible nature of our problems that protects them from solutions. Part of this is due to our hiding from them in view of our lack of any alternatives. But if we look to those four concepts dipping into emotions—"cultural values," "relative deprivation," "reinforcement," and "alienation"—we can understand how amorphous and intangible they are relative to, say, concepts dealing with social organization like social stratification, bureaucracy and group. Figure 1-3 helps us to make such phenomena visible not only by giving them names but also by its systematic approach. If those names emphasize the denotation of those concepts, then that systematic approach emphasizes their connotation. This is of course basic to our web orientation to the scientific method: presenting a system of ideas, versus seeing ideas in isolation.

We might also pay attention to Chapter 1's presentation of the nature of the scientific method as we proceed with our focus on emotions. As a first and perhaps the most important step of that method, we have centered on the definition of a problem. This may sound trite, yet it is not. Defining trivial problems, such as those relevant only to one of our forty sections of the ASA, is a common means of avoiding any deep emotional involvement in the problem. And once a problem is defined in this way the rest of the research process rarely alters that definition. By contrast, our web approach sees all phenomena in interaction with one another, just as Figure 1-3 shows such linkages. This being the case, all substantive and applied problems also interact with one another. As a result, *no* problem can remain restricted to one particular segment of society. For example, someone's prejudice against members of a minority group linked to feelings of **relative deprivation** derives its force in part from widespread patterns of **social stratification,** which tend to structure those hierarchical feelings. Further, feelings of **alienation** would probably play a role here as well, since this **structure** within the individual emphasizes **negative reinforcement.** Historically, all of this is encouraged by the outer-oriented **worldview** and **anomie** to be found throughout modern society. Prejudice located within one particular scene, then, cannot be separated from a wide range of social problems.

Our definition of the scientific method also includes a reflexive orientation, an approach we elaborated on in Chapter 5. This quote from my own efforts to develop such an orientation illustrates the emotional dimension invoked here:

> To illustrate very briefly what a reflexive analysis of the research situation might involve, I might refer to my own hesitations, fears and shame relative to the research problem of the accelerating gap between aspirations and fulfillment sketched in this chapter. Where do I get the chutzpah to dare to point toward an alternative paradigm for the entire discipline of sociology? And far beyond this, who am I to propose nothing less than a change in our worldview if we are to achieve that alternative paradigm? . . . Such hesitation, fear and shame manifest themselves in many ways, such as burying my writing in endless qualifications, hiding behind the statements of a great many other sociologists and philosophers, writer's block, repetitive material, numerous drafts and overly intellectualized writing.

A reflexive analysis would raise to the surface emotions like fear and shame. It would challenge our outer-oriented worldview within which our bureaucratic scientific paradigm appears to be nested. And it would open up the possibility of shifting to a worldview that supports an interactive approach to the scientific method. Granting that such a challenge to our worldview bites off a great deal, it nevertheless points in an essential direction.

Part Two: Illustrating the Web Approach to the Scientific Method

Our discussion of **anomie** in Chapter 3 brings to the fore once again the enormous frustrations implied in Figure 1-1, deriving from our inability to fulfill the "revolution of rising expectations" within modern society. Although Durkheim did not focus on such emotional problems directly, he catalogued their resultant in increasing suicide rates as modernization proceeded. Weber (1970) saw those escalating **cultural values**—labeled by Williams as "achievement and success," "activity and work," [material] "progress," and "efficiency and practicality"—as deriving in part from the Protestant ethic and functioning to motivate our continuing industrial revolution. Karen Horney, in common with Durkheim, analyzed modern society's basic contradictions. She went on to describe their negative impact on the individual in her *The Neurotic Personality of Our Time* (1937), such as the contradiction "between the stimulation of our needs and our factual frustrations in satisfying them." In this way she was able to link cultural values and their widespread lack of fulfillment within social structure, on the one hand, with the structure of the **individual**. For Horney, the crucial source of the individual's neurotic conflicts lies within the cultural contradictions of modern society. To understand the source of our emotional problems, then, it is not sufficient to look to biology or to the individual's family, as so many analysts and psychotherapists maintain. We must also look to the culture of modern society.

Alienation is the sociological concept most directly tied to the emotional state of the modern individual. In *The Languages of Pao* we can see that state dramatized by the culture of the planet Breakness, where each individual is a world unto himself and where there are no concepts for emotions. Granting Marx's paramount emphasis on social organization, he defined alienation broadly enough so as to include the individual's **biological structure** and relation to **physical structures.** By so doing he forged links— just as Horney did a century later—between the social structure of modern society and the emotional problems of the individual. Weber and Simmel joined him in seeing the individual as being crushed by the forces of modernity. Weber saw modernization as yielding "formal rationality" but not **substantive rationality,** creating "specialists without spirit, sensualists without heart." Simmel saw the maintenance of individuality as associated with "the deepest problems of modern life":

> [The individual] is reduced to a negligible quantity. He becomes a single cog in the vast overwhelming organization of things and forces which gradually take out of his hands everything connected with progress, spirituality and value. ([1903] 1971:337)

Simmel sees modern society as depriving the **individual** of progress—implying substantive rationality—spirituality and value.

Moving from alienation to **social stratification** within Chapter 3, we can follow Gramsci's analysis of "hegemony" and see stratification within **culture** no less than within social organization, taking stratification closer to the emotions of the individual. In addition to illustrating social stratification using rankings of occupations, education, and income, we can— following Foucault—look to the **domination** of an audience by an expert. LaFountain—following Foucault's approach—has analyzed Dr. Ruth Westheimer's approach to sexuality in her radio and television programs. He finds her promulgating a "repressive hypothesis," which yields domination in the name of liberation. On the one hand, she speaks out against the silencing, censoring, and repression of sex due to "scrupulousness, an overly acute sense of sin," and "hypocrisy." Yet on the other hand her impact is much different:

What appears as liberating is arguably little more than the promotion of a celebrity and a media form at the expense of those who become dependent on her for perspective and renewal. (LaFountain 1989:135)

Dr. Ruth achieves domination over her audience within her media scenes. Since such domination is linked to widespread patterns of social stratification throughout modern society, we can come to see Dr. Ruth as mouthing the importance of sexual and emotional expression but in fact reinforcing sexual and emotional repression.

As for **relative deprivation**, that sociological concept along with just about every other one has been sidetracked so as to be associated with a narrow range of phenomena, versus any effort to see it in a very abstract or general way. It is viewed as one possible predictor of political revolutions, yet such predictive efforts generally take us away from achieving broader understanding, and that achievement is exactly what this concept promises. It is located within the emotional area, it is a situational concept, and it links directly with patterns of social stratification. By remaining unaware of the situational phenomenon of relative deprivation, our approach to the structural phenomenon of social stratification remains static. By combining the two, however, we can begin to understand both how stratification makes its presence felt in any given scene and also just how our emotions come into play in one scene after another. Yet we need not limit ourselves to an understanding of how our present bureaucratic cultural paradigm works; we can also gain insight into how to change it. For example, relative deprivation suggests a seesaw metaphor or **image of the situation** conveying a bureaucratic worldview. But we can develop a different **image of the situation,** a stairway, which invokes an interactive worldview. And, emotionally, instead of gaining positive reinforcement from playing a seesaw game, the individual can learn to gain positive reinforcement by playing a stairway game.

Whereas Chapter 3 of Part II emphasizes problems, Chapter 4 centers on directions for solutions with a focus on the invisible crisis of modern society. The initial section on a general understanding of the role of interaction in the crisis includes discussion of interaction among the rows and among the columns of Figure 1-3. Those rows and columns are delineated and illustrated more clearly in Figures 5-1 and 6-1. Such interaction implies that, metaphorically, "head," "heart," and "hand" are all operating simultaneously all of the time. Our emotions, then, appear to be involved in every instance of our behavior. Carrying this analysis a bit further, to the extent that we are located within a bureaucratic cultural paradigm and worldview, those emotions will parallel the patterns of social stratification associated with that worldview. To illustrate, many emotions appear to be oriented hierarchically. If we use the seesaw as our image of the situation, then feelings of fear, guilt, envy, shame, and pessimism place us at the bottom of the seesaw relative to others. By contrast, emotions such as anger, hate, haughtiness, arrogance, and disdain place us at the top of the seesaw relative to others. Following our analysis, we feel such emotions all the time, whether or not we are aware of them or express them in an overt manner. And since they are generally viewed as undesirable, we tend to repress them. Yet concepts like **image of the situation** and **social stratification** can help us to make those emotions more visible and, as a result, help us to change them.

Gandhi's approach to conflict resolution or *satyagraha* illustrates the central role of emotions as a tool for solving fundamental problems. If we look to the eight elements of *satyagraha* as interpreted within the literature of conflict resolution, we find all of them pointing in the direction of an interactive versus a bureaucratic worldview, with half of them singling out emotions in particular. For example, (3) states that the rival group is not to be *humiliated,* (6) suggests the importance of *friendly* verbal discussion with the rival group, (7) suggests developing a consistent attitude of *trust* toward them, and (8) suggests *empathy* with respect to their "motives, affects, expectations" and "attitudes," all of which have to do with emotions. This approach to emotions is an effort to achieve consistency between means and ends, just as Gandhi illustrated that consistency in his own approach to *satyagraha.* If the end is an India no longer ruled from abroad but able to interact with other states as an independent state, then that end points away from social stratification as a means and toward egalitarian relationships. Such consistency implies a worldview within which the range of situational and structural factors involved in relations with the rival group— as depicted in Figure 1-3 and Figure 6-2—all point in that same direction. We can see further illustrations of this when we examine the remaining four elements of *satyagraha,* with their emphasis on egalitarian interaction and avoiding violence.

Scheff's analysis at the end of Chapter 4 centers on emotions, seeing attention to emotions as a basis for solving problems. He and Retzinger quote a mediator's response in a custody battle where the wife just declared to her ex-husband, "You never paid any attention to the children, then you left me, and you're not getting the children now or ever":

> The anger and hurt you feel right now is not unusual, and it is very understandable. It is also not unusual for a parent who was not involved with the children before a divorce to decide to become sincerely involved after the divorce. Allowing that opportunity will give your children a chance to get to know their father in the future in a way that you wanted in the past. But give yourself plenty of time to get through these difficult feelings. (Retzinger and Scheff 2000)

Here, the mediator expresses the anger, hurt, and shame of the parents, but in a way that points toward a stairway **image of the situation.** This contrasts with the seesaw image just depicted by the mother, which would have produced a corresponding reaction by the father. Retzinger and Scheff applaud the mediator for avoiding the "feeling trap"—a causal loop—which would have made it difficult for the couple to remain coparents.

It is one thing for a mediator to help us recognize our feelings and bring them out into the open as a basis for solving interpersonal problems, but how is the individual to learn to do this without such help? More generally, how are we moderns to learn to achieve the balance among "head," "heart," and "hand" that appears to have been the case in general for preliterate society prior to our emphasis on the "head" at the expense of the "heart"? And if our repressed emotions like shame are there from one moment to the next yet remain invisible, how are we to learn to make them visible? How can we moderns move "back to the future" by returning to the preliterate balance, given an orientation to language and culture much like that of those on the planet Breakness? Our analysis points toward nothing less than changing our language and culture, following the experiment with the languages of Pao. Instead of learning Pastiche we can be guided by a web approach to the scientific method. Instead of embarking on a twenty-year experiment with young children to change our culture, we can all learn to be guided by an interactive cultural paradigm and worldview. And instead of developing specialized enclaves that embody this approach to language and culture, we can follow Beran Panasper's vision—much like that of Mills and Gouldner—of teaching everyone the new language and culture.

Part Three: Some Implications

Gouldner's reflexive idea puts forward an enormous challenge, since it flies in the face of our cultural paradigm and worldview. Let us quote once again his idea about transcending ordinary language:

> The pursuit of hermeneutic understanding, however, cannot promise that men as we now find them, with their everyday language and understanding, will always be capable of further understanding and of liberating themselves. At decisive points the ordinary language and conventional understandings fail and must be transcended. It is essentially the task of the social sciences, more generally, to create new and "extraordinary" languages, to help men learn to speak them, and to mediate between the deficient understandings of ordinary language and the different and liberating perspectives of the extraordinary languages of social theory. (1972:16)

What appears to be crucial here is that the sociologist not think of himself or herself as superior to "men as we now find them" but rather point toward learning, along with others, "the different and liberating perspectives of social theory." It is by no means enough for the sociologist to develop ideas about the language and emotions of others: he or she must also probe self. Yet granting the importance of such opposition to a worldview that appears to yield fundamental and increasing problems, the difficulties involved stagger the imagination.

What would constitute highly effective technologies for helping to fulfill Gouldner's vision of a reflexive sociology? In Chapter 5 three examples are cited: audiovisual experiences as devices for learning how to apply sociological concepts, a diary as a way of applying those concepts to oneself, and the writing of a book. These all appear to be useful procedures, yet can we in addition manage to harness the full power of electronic gaming with all its potential excitement and computer technology with all its computational possibilities so as to work in the same direction? For example, electronic games help players to feel very good about their ability to destroy countless enemies, following a bureaucratic or seesaw **image of the situation.** Can electronic games be invented that help us to see emotions that otherwise would remain invisible? And can those same games help us to invoke emotions like love and confidence rather than those of hate and fear? Can computer technology be developed to predict the outcomes of situations based on inputs of the concepts in Figure 1-3? And can the individual, by plugging in such inputs, learn to understand the complexity of ordinary situations? What promises that such gaming and computer technologies can be constructed is an approach to the scientific method that can integrate that range of concepts and apply them systematically to a given scene. And what promises to attract investment in the development of such gaming and computer technologies is the possibility that instead of feeding the destructive orientations within our seesaw worldview, they can help us learn how to express our emotions and thus develop more fully as individuals.

Figure 5-1, later bolstered by Figure 6-1, points us away from the requirement that everyone must learn to use the language of sociology in order to shift scientific and cultural paradigms. This figure gives us a series of bases to touch if we wish to open up to the complexity of a given scene. Three of the nine bases—those headed "Heart"—have to do directly or indirectly with our emotions. And since all nine categories can be applied to any given scene, our emotions become an essential aspect of any and all of the situations in which we find ourselves. To understand the significance of this figure we might metaphorically see ourselves as individuals using the language of the planet Breakness, where we have no words whatsoever for any of our emotions. This is an exaggeration, but it nevertheless is not far from the situations in which we Springdalers find ourselves, having learned to repress emotions that otherwise would interfere with our ability to attain "some degree of satisfaction, recognition and achievement." Assuming such repression to be a modern problem, Figure 5-1 gives us a direction for addressing it: taking into account the heart—in relation to social structures, situations, and individual structures—in attempting to understand any given situation. And beyond that understanding, Figure 5-1 suggests that, like the wizards of the planet Breakness, we will always remain warped human beings unless we learn how to freely express our emotions in everyday life.

Moving to Chapter 6, Figure 6-1 presents new concepts that emphasize an interactive worldview by contrast with concepts emphasizing a bureaucratic worldview in Figure 1-3. These new concepts provide a better basis for an alternative to that bureaucratic worldview, without which it would be difficult or impossible for the individual to question the latter seriously. To illustrate with respect to the structure of the individual, we have three new concepts that, together, point toward a balanced head-heart-hand orientation: **imaginative orientation, expressive orientation,** and **praxis orientation.** "Expressive orientation" contrasts with the concept within the same category of Figure 6-1 previously emphasized: **alienation.** If "alienation" helps us to understand modern problems, then "expressive orientation" suggests modern solutions. We can come to see these three concepts as forming a positive loop within a causal-loop diagram, a loop that leads to accelerating movement toward an interactive cultural paradigm where head, heart, and hand are balanced. The imaginative orientation can yield an image of a stairway world where—with the aid of an expressive orientation—the individual can develop deeper commitments or motivation for activities seen as more meaningful. A praxis orientation would yield effective actions on the basis of such commitments, encouraging in turn further movement around the loop and further developing all three orientations.

SOME CONCLUDING REMARKS

Looking back over our back-to-the-future journey in this chapter with the aid of new concepts, causal-loop and metaphorical tools, are we in fact following too closely in the path of the Springdalers with their techniques of particularization and falsification of memory? Have we failed to pay serious attention to what it would take to change not only our emotional lives but also to change the way we live from one moment to the next? And in this book as a whole, have we held so closely to the Enlightenment dream for sociology, social science, and society—ignoring evidence against its feasibility along with memories of the Holocaust and other twentieth-century horrors forged by human beings—as to depart, like the Springdalers, from the plane of reality? We can look back with admiration to Einstein's ability to develop somehow a new scientific paradigm for physics despite Newton's hold on his own mind and that of others. And we can also admire his persistence despite opposition from many quarters and despite a way of life in Europe and elsewhere that was shattering into tiny pieces. Yet what we are now facing if we are to give credibility to the preceding arguments—and I include myself here as well—is a problem of far greater magnitude. It is nothing less than challenging the only game in town, our worldview and cultural paradigm, from which we derive the meaning in our lives from one moment to the next. To what extent have this chapter and the previous ones failed, in Springdaler fashion, to open up to the enormous problems we presently are facing?

A recent news story can help bring us down to the plane of reality. David F. Gordon, national intelligence officer for economics and global issues at the CIA's National Intelligence Council, informed a House committee of "an increasing possibility of a biological terrorist attack against the United States, and that the attack could be mounted through an infectious disease." Such diseases—such as AIDS—have "steadily become more drug-resistant because of underuse of antibiotics in poor countries and overuse in wealthier ones," underlining President Clinton's act earlier in the year declaring HIV and AIDS a national security threat and his earlier warnings on threats from terrorism. Gordon went on to imply the relative invisibility of this danger:

> If this were a military invasion, it would be easier to galvanize the public and policy-makers. . . . For humanitarian and self-survival reasons, we need to face this challenge as if it were an invading army. (Donnelly, 2000:A1)

Jack Vance wrote in his *The Languages of Pao* about a twenty-year period during which new languages could take hold and alter the way of life of Pao society dramatically. How do we face this threat—tied as it is, following the arguments in this book, to every invisible and visible modern social problem—in the time that we have left?

Postmodernists among others have taught us that we have no guarantees that social science can save us, yet we need no guarantees to look to an approach—with the track record for rapid cumulative development and problem-solving of the scientific method—that appears to carry with it the potential of fulfilling our Enlightenment dream. Given the urgent problems we moderns face, and given the relative invisibility of many of them, we may well fail in the attempt. Yet that attempt is our best hope not merely for fulfilling that dream but also for our own survival and that of the generations that may or may not follow us. In this book we have tried to reconstruct the scientific method, following Kuhn's implications, so that we can employ it to pull together the social science knowledge we presently have in bits and pieces and use that knowledge to construct an alternative paradigm for sociology and society, since *some* alternative is required if we are to criticize our present paradigms. Our key argument here is for a reconstructed scientific method, granting that such a method may subsequently yield knowledge that overturns the initial knowledge based on using that method, which is presented here. Perhaps we will learn that matters are not quite so urgent as what is depicted here. Or perhaps we will learn that matters are in fact far worse. Yet it is on a reconstructed scientific method that we pin our hopes for fulfilling that Enlightenment dream and finding some way out of our problems.

If we follow the arguments in this book, then it appears that we sociologists are saddled with an incredible responsibility for the future of society. It is we who have the greatest possibility, based on the achievements of both classical and modern sociologists and following Gouldner, "to create new and 'extraordinary' languages, to help men learn to speak them, and to mediate between the deficient understandings of ordinary language and the different and liberating perspectives of the extraordinary languages of social theory." And far beyond the problems now faced by modern society, we have the potential to help create a world where every individual can learn—following Mills—to develop a sociological imagination or, more specifically, an **imaginative, expressive,** and **praxis orientation.** We can predict the nature of that world no more than preliterates could have predicted the nature of modern society, although our analysis of an interactive cultural paradigm and worldview in this book provides hints. Yet in order for us sociologists to help create that world, it appears that we must follow the ideas and ideals, if not the actions, of Gouldner and Mills. We must learn to move out from our expert and alienated society in our everyday and professional lives and toward a society that increasingly fulfills the extraordinary potential of all human beings.

Glossary

action: individual behavior

addiction: the individual's subordination of individuality to dependence on external phenomena

alienation: persisting feelings of isolation from self, others, one's own biological structure, and the physical universe

anomie: the failure of society's norms or rules to guide the individual's actions toward the fulfillment of values or interests

biological structure: a system of elements that interacts to a relatively great extent with its environment

bureaucracy: a group with limited yet persistent interaction up and down its hierarchy and across its specialized fields

conformity: legitimate behavior as defined by norms and values for a given situation

culture: the widely shared interests and beliefs of a people that (1) are learned with the aid of language and persist, and (2) shape and are shaped by people's momentary behavior

definition of the situation: the individual's understanding of the momentary scene

deviance: illegitimate behavior as defined by norms and values for a given situation

expressive orientation: the individual's commitment to awareness and expression of emotions

group: a collection of individuals who share certain characteristics

image of the situation: the individual's view of the momentary scene

imaginative orientation: the individual's openness to learning

individual: a system of social, personality, biological, and physical structures

institutions: systems of norms and values centered on solving a given problem throughout society

interaction: individual action that yields environmental response

label: assigning an individual or group to a given linguistic category within a particular situation

norms: shared beliefs or expectations within a group

personality structure: the individual's patterns of action, interaction, interests, and beliefs

physical structure: a system of elements that interacts to a relatively small extent with its environment

praxis orientation: the individual's commitment to interaction that shapes self and world

reinforcement: the fulfillment of the individual's interests, motives or needs within a given situation

relative deprivation: the individual's feeling of unjustified loss or frustration of value fulfillment relative to others who are seen as enjoying greater fulfillment

scientific method: a procedure for achieving deepening understanding of problems that builds on prior knowledge and is gained through patterns of social interaction

scientific technology: a procedure for solving problems that builds on prior knowledge and is gained through patterns of social interaction

self-image: the individual's view of self

situation: any phenomenon located in time and space

social interaction: momentary action that mutually affects two or more individuals and encompasses a given range of phenomena

social organization: persisting and shared patterns of action or interaction

social stratification: a persisting hierarchy or pattern of inequality within a group

social structure: persisting and shared patterns of action, interaction, interests, and beliefs

socialization: a learning process where the individual develops a personality and culture is transmitted

structure: a persisting system of elements

substantive rationality: the individual's orientation to the full range of cultural values

values: shared interests or ideals within a group

worldview: The individual's *Weltanschauung* or global outlook that is widely shared throughout society

References

Apel, Karl. 1980. *Toward a Transformation of Philosophy*. London: Routledge.

Bailey, F. G. 1977. *Morality and Expediency*. Oxford: Blackwell.

Ball-Rokeach, Sandra J., Milton Rokeach, and Joel W. Grube. 1984. *The Great American Values Test*. New York: Free Press.

Banks, Jane, and Jonathan David Tankel. 1990. "Science asFiction: Technology in Prime Time Television." *Critical Studies in Mass Communication* 7:24–36.

Bell, Daniel. 1973. *The Coming of Post-Industrial Society*. New York: Basic Books.

Bell, Wendell. 1957. "Anomie, Social Isolation, and the Class Structure." *Sociometry* 20(June):105–16.

Benedict, Ruth. 1946. *The Chrysanthemum and the Sword*. Boston: Houghton-Mifflin.

Bergesen, Albert. 1993. "The Rise of Semiotic Marxism." *Sociological Perspectives* 36:1–22.

Blalock, Hubert M., Jr. 1984. *Basic Dilemmas in the Social Sciences*. Beverly Hills: Sage.

Blau, Peter M. 1964. *Exchange and Power in Social Life*. New York: Wiley.

Blauner, Robert. 1964. *Alienation and Freedom*. Chicago: University of Chicago Press.

Blumer, Herbert. 1956. "Sociological Analysis and the 'Variable,'" *American Sociological Review* 21(December):683–90.

Blumer, Herbert. 1969. *Symbolic Interactionism: Perspective and Method*. Englewood Cliffs, NJ: Prentice-Hall.

Bondurant, Joan V. 1965. *Conquest of Violence*. Berkeley: University of California Press.

Bowles, Samuel, and Herbert Gintis. 1987. *Democracy and Capitalism*. New York: Basic Books.

Braverman, Harry. 1974. *Labor and Monopoly Capital*. New York: Monthly Review Press.

Brinton, Crane. 1952. *The Anatomy of Revolution*. Englewood Cliffs, NJ: Prentice-Hall.

Brown, Donald E. 1991. *Human Universals*. New York: McGraw-Hill.

Burawoy, Michael. 1979. *Manufacturing Consent*. Chicago: University of Chicago Press.

Chapin, F. Stuart. 1952. *Social Participation Scale*. Minneapolis: University of Minnesota Press.

Chasin, Barbara H. 1990. "C. Wright Mills, Pessimistic Radical." *Sociological Inquiry* 60(Fall):337–51.

Collins, Randall. 1975. *Conflict Sociology*. New York: Academic Press.

Collins, Randall. 1998. "The Sociological Eye and Its Blinders." *Contemporary Sociology* 27:2–7.

Constas, Helen. 1958. "Max Weber's Two Conceptions of Bureaucracy." *American Journal of Sociology* 52(January):400–9.

Cooley, Charles H. 1922. *Human Nature and the Social Order*. New York: Scribner's.

Cronbach, Lee J., and Paul E. Meehl. 1955 "Construct Validity in Psychological Tests." *Psychological Bulletin* 52(May):281–302.

Crutchfield, Robert D. 1992. "Anomie and Alienation." In Edgar F. Borgatta (Ed.), *Encyclopedia of Sociology* (pp. 95–100). New York: Macmillan.

Dahrendorf, Ralf. 1959. *Class and Class Conflict in Industrial Society*. Stanford, CA: Stanford University Press.

Davies, James C. 1962. "Toward a Theory of Revolution." *American Sociological Review* 27(February):5–19.

Davies, James C. (Ed.). 1971. *When Men Revolt and Why*. New York: Free Press.

Davis, Kingsley and Moore, Wilbert E. 1945. "Some Principles of Stratification." *American Sociological Review*10:242–49.

Derber, Charles, William A. Schwartz, and Yale Magrass. 1990. *Power in the Highest Degree*. New York: Oxford.

Dewey, John. [1920] 1948. *Reconstruction in Philosophy*. Boston: Beacon.

Diesing, Paul. 1991. *How Does Social Science Work?* Pittsburgh: University of Pittsburgh Press.

Donnelly, John. 2000. "CIA Sees Threat in Global Diseases." *Boston Globe*, June 30, A1.

Duhem, Pierre. 1954. *The Aim and Structure of Physical Theory*. Princeton, NJ: Princeton University Press.

Durkheim, Emile. [1897] 1951. *Suicide: A Study in Sociology* (translated by John A. Spaulding and George Simpson). Glencoe, IL: Free Press

Edin, Kathryn, and Laura Lein. 1997. *Making Ends Meet: How Single Mothers Survive Welfare and Low-Wage Work*. New York: Russell Sage Foundation.

Edwards, Richard. 1979. *Contested Terrain*. New York: Basic Books.

Elias, Norbert. 1978. *The History of Manners*. New York: Pantheon.

Emerson, Richard M. 1962. "Power-Dependence Relations." *American Sociological Review* 27:31–41.

Emerson, Richard M. 1976. "Social Exchange Theory." *Annual Review of Sociology* 2:335–62.

Erikson, E. 1986. "On Work and Alienation." *American Sociological Review* 51(February):1–8.

Etzioni-Halevy, Eva (Ed.). 1997. *Classes and Elites in Democracy and Democratization*. New York: Garland.

Farson, Richard E. 1977. "Why Good Marriages Fail." In Ronald Fernandez (Ed.), *The Future as a Social Problem* (pp. 169–76). Santa Monica, CA: Goodyear.

Feldberg, Roslyn L., and Evelyn Nakano Glenn. 1983. "Technology and Work Degradation." In Joan Rothschild (Ed.), *Machina Ex Dea*. New York: Pergamon.

Feynman, Richard. "Cargo Cult Science." In *Surely You're Joking, Mr. Feynman* (pp. 308–18). New York: Bantam.

Forrester, Jay W. 1968. *Principles of Systems*. Cambridge, MA: MIT Press.

Forrester, Jay W. 1969. *Urban Dynamics*. Cambridge, MA: MIT Press.

Forrester, Jay W. 1971. *World Dynamics*. Cambridge, MA: Wright-Allen.

Foucault, Michel. 1972. *The Archeology of Knowledge*. New York: Harper & Row.

Foucault, Michel. 1977. *Madness and Civilization*. London: Tavistock.

Foucault, Michel. 1978. *The History of Sexuality*, Vol. 1: *An Introduction*. New York: Pantheon.

Foucault, Michel. 1984a. The History of Sexuality, Vol. 2: *The Uses of Pleasure*. Paris: Editions Gallimard.

Foucault, Michel. 1984b. *The History of Sexuality*, Vol. 3: *The Care of the Self*. Paris: Editions Gallimard.

Freeman, Derek. 1983. *Margaret Mead and Samoa*. Cambridge, MA: Harvard University Press.

Freeman, Derek. 1989. "Fa'apua'a Fa'amu and Margaret Mead, *American Anthropologist* 91:1017–22.

Freire, Paulo. 1970. *Pedagogy of the Oppressed*. New York: Herder and Herder.

Freire, Paulo. 1973. *Education for Critical Consciousness*. New York: Seabury.

Gadamer, Hans. [1960] 1975. *Truth and Method*. New York: Seabury.

Geertz, Clifford. 1973. *The Interpretation of Cultures*. New York: Basic Books.

Geertz, Clifford. 1983. *Local Knowledge*. New York: Basic Books.

Glenn, Evelyn Nakano, and Roslyn L. Feldberg. 1977. "Degraded and Deskilled." *Social Problems* 25:52–64.

Goffman, Erving. 1959. *The Presentation of Self in Everyday Life*. Garden City, NY: Anchor.

Goffman, Erving. 1967. *Interaction Ritual: Essays on Face-to-Face Behavior*. Garden City, NY: Anchor.

Goldstone, Jack A. 1982. "The Comparative and Historical Study of Revolutions." *Annual Review of Sociology* 8:187–207.

Gouldner, Alvin W. 1970. *The Coming Crisis of Western Sociology*. New York: Basic Books.

Gouldner, Alvin W. 1972. "The Politics of the Mind: Reflections on Flack's Review of *The Coming Crisis of Western Sociology*." *Social Policy* 5(March / April):13–21, 54–58.

Gramsci, Antonio. 1971. *Selections from the Prison Notebooks of Antonio Gramsci*, edited and translated by Quintin Hoare and Geoffrey Nowell-Smith. London: Lawrence and Wishart.

Granovetter, Mark S. 1985. "Economic Action, Social Structure, and Embeddedness." *American Journal of Sociology* 91:481–510.

Greenstein, Theodore N. 1989 "Modifying Beliefs and Behavior through Self-Confrontation." *Sociological Inquiry* 59(November):396–407.

Gurr, Ted. 1968. "A Causal Model of Civil Strife." *American Political Science Review* 62(December):1104–24.

Gurr, Ted. 1970. *Why Men Rebel*. Princeton, NJ: Princeton University Press.

Gusfield, Joseph R. 1968, "The Study of Social Movements." *International Encyclopedia of the Social Sciences* (445–52). New York: Macmillan.

Habermas, Jurgen. 1971. *Knowledge and Human Interests*. Boston: Beacon Press.

Hayakawa, Samuel I. 1949. *Language in Thought and Action*. New York: Harcourt, Brace & World.

Hempel, Carl G. 1965. *Aspects of Scientific Explanation*. New York: Free Press.

Henshel, Richard L. 1982a. "Sociology and Social Forecasting." *Annual Review of Sociology* 8:57–79.

Henshel, Richard L. 1982b. "The Boundary of the Self-Fulfilling Prophecy and the Dilemma of Social Prediction." *British Journal of Sociology* 33(December):511–28.

Henshel, Richard L. 1990. "Credibility and Confidence Loops in Social Prediction." In Felix Geyer and Johannes Van Der Zouwen (Eds.), *Self-Referencing in Social Systems*. New York: Intersystems.

Henshel, Richard L. (1993) "Do Self-Fulfilling Prophecies Improve or Degrade Predictive Accuracy?" *Journal of Socio-Economics* 22:85–104.

Henshel, Richard L., and William Johnston. 1987. "The Emergence of Bandwagon Effects." *Sociological Quarterly* 28:493–511.

Himmelstein, Jerome L., and Kimmel, Michael S. 1981. "Review Essay: States and Revolutions." *American Journal of Sociology* 86:1145–54.

Holton, Gerald. 1996. *Einstein, History and Other Passions: The Rebellion Against Science at the End of the Twentieth Century*. Reading, MA: Addison-Wesley.

Homans, George C. 1964. "Bringing Men Back In." *American Sociological Review* 29(December):809–18.

Homans, George C. 1958. "Social Behavior as Exchange." *American Journal of Sociology* 62:597–606.

Horney, Karen. 1937. *The Neurotic Personality of Our Time*. New York: Norton.

Hornig, Susanna. 1990. "Television's NOVA and the Construction of Scientific Truth." *Critical Studies in Mass Communication* 7:11–23.

Horowitz, Irving L. 1983. *C. Wright Mills: An American Utopian*. New York: Free Press.

Hyman, Herbert H. 1972. *Secondary Analysis in Sample Surveys*. New York: Wiley.

Illich, Ivan. 1972. *Deschooling Society*. New York: Harper & Row.

Janis, Irving L., and Daniel Katz. 1959. "The Reduction of Intergroup Hostility." *Journal of Conflict Resolution* 3(March):85–100.

Jerusalem Bible: Reader's Edition. 1966. Garden City, NY: Doubleday.

Johnston, Douglas, and Cynthia Sampson (Eds.). 1994. *Religion, The Missing Dimension of Statecraft*. New York: Oxford.

Jones, Richard Foster. 1965. *Ancients and Moderns: A Study of the Rise of the Scientific Movement in Seventeenth Century England*. Berkeley: University of California Press.

Kaplan, Abraham. 1964. *The Conduct of Inquiry*. San Francisco: Chandler.

Kapuscinski, Ryszard. 1985, "Revolution." *New Yorker*, March 11, 86–101.

Kincaid, Harold. 1996. *Philosophical Foundations of the Social Sciences*. New York: Cambridge University Press.

Kincaid, Harold. 2000. "Formal Rationality and Its Pernicious Effects on the Social Sciences." *Philosophy of the Social Sciences* (March):67–88.

Kohn, Melvin L. 1976. "Occupational Structure and Alienation." *American Journal of Sociology* 82:111–30.

Kohn, Melvin L., and Carmi Schooler. 1983. *Work and Personality*. Ablex.

Korzybski, Alfred. 1933. *Science and Sanity*. Garden City, NY: Country Life.

Kuhn, Thomas S. 1962. *The Structure of Scientific Revolutions*. Chicago: University of Chicago Press.

Kuhn, Thomas S. 1977. *The Essential Tension*. Chicago: University of Chicago Press.

Kuhn, Thomas S. 1992. "The Trouble with the Historical Philosophy of Science." Department of the History of Science, Harvard University, Cambridge, MA.

LaFountain, Marc J. 1989. "Foucault and Dr. Ruth." *Critical Studies in Mass Communication* 6:123–37.

Latour, Bruno. 1987. *Science in Action.* Cambridge, MA: Harvard University Press.

Lauderdale, Pat, Steve McLaughlin, and Annamarie Oliverio. 1990. "Levels of Analysis, Theoretical Orientations and Degrees of Abstraction." *American Sociologist* 21(Spring):29–40.

Lazarsfeld, Paul, and Morris Rosenberg (Eds.). 1955. *The Language of Social Research.* New York: Free Press.

Lenski, Gerhard E. 1966. *Power and Privilege.* New York: McGraw-Hill.

Lerner, Daniel. 1958. *The Passing of Traditional Society.* New York: Free Press.

Levin, Jack. 1968. *The Influence of Social Frame of Reference for Goal Fulfillment on Social Aggression.* Ph.D. dissertation, Boston University.

Levin, Jack. 1975. *The Functions of Prejudice.* New York: Harper & Row.

Lewis, Helen. 1971. *Shame and Guilt in Neurosis.* New York: International Universities.

Lipset, Seymour Martin, and Earl Raab. 1978. *The Politics of Unreason.* Chicago: University of Chicago Press.

Lo, Clarence Y. H. 1992. "Alienation." In Edgar F. Borgatta (Ed.), *Encyclopedia of Sociology* (pp. 48–54). New York: Macmillan. 1992.

Lowenthal, Leo. 1956. "Biographies in Popular Magazines." In W. Petersen (Ed.), *American Social Patterns* (pp. 63–118). New York: Doubleday.

Lundberg, George A. 1961. *Can Science Save Us?* New York: McKay.

Luria, Aleksandr R. 1976. *Cognitive Development* (Michael Cole, editor). Cambridge, MA: Harvard University Press.

Malotki, Ekkehart. 1983. *Hopi Time.* Berlin: Mouton.

Mannheim, Karl. [1921–1922] 1952. "On the Interpretation of *Weltanschauung.*" In *Essays on the Sociology of Knowledge.* New York: Oxford University Press.

Maruyama, Magoroh. 1963. "The Second Cybernetics." *American Scientist* 51:164–79.

Marx, Karl. [1844] 1964. *Early Writings.* (T. B. Bottomore, translator and editor). New York: McGraw-Hill.

Mead, Margaret. 1928. *Coming of Age in Samoa.* New York: Morrow.

Meadows, Donella H. 1980. "The Unavoidable A Priori." In Jorgen Randers (Ed.), *Elements of the System Dynamics Method.* Cambridge, MA: MIT Press.

Meadows, Donella H., et al. 1972. *The Limits to Growth.* New York: Universe.

Meier, Dorothy L., and Wendell Bell. 1959. "Anomie and Differential Access to the Achievement of Life Goals." *American Sociological Review* 24(April):189–202.

Merton, Robert K. 1936. "The Unanticipated Consequences of Purposive Social Action." *American Sociological Review* 1(December):894–904.

Merton, Robert K. [1938] 1996. "The Rise of Modern Science." In *On Social Structure and Science* (Piotr Sztompka, editor). Chicago: University of Chicago Press.

Merton, Robert K. 1949. "Social Structure and Anomie." In *Social Theory and Social Structure* (pp. 125–49). New York: Free Press.

Merton, Robert K. 1968. *Social Theory and Social Structure.* New York: Free Press.

Meszaros, Istvan. 1970. *Marx's Theory of Alienation.* New York: Harper & Row.

Michels, Robert. 1949. *Political Parties*. New York: Free Press.

Mills, C. Wright. 1940. "Situated Actions and Vocabularies of Motive." *American Sociological Review* 5:904–13.

Mills, C. Wright. 1943. "The Professional Ideology of Social Pathologists." *American Journal of Sociology* 49(September):165–80.

Mills, C. Wright. 1948. *The New Men of Power*. New York: A. M. Kelley.

Mills, C. Wright. 1951. *White Collar*. New York: Oxford University Press.

Mills, C. Wright. 1956. *The Power Elite*. New York: Oxford University Press.

Mills, C. Wright. 1958. *The Causes of World War Three*. New York: Simon & Schuster.

Mills, C. Wright. 1959. *The Sociological Imagination*. New York: Oxford University Press.

Mizruchi, Ephraim Harold. 1960. "Social Structure and Anomia in a Small City." *American Sociological Review* 25(October):645–54.

Moberg, David O. 1962. *The Church as a Social Institution*. Englewood Cliffs, NJ: Prentice-Hall.

Molm, Linda D., and Karen S. Cook. 1995. "Social Exchange and Exchange Networks." In Karen S. Cook, Gary Alan Fine, and James S. House (Eds.), *Sociological Perspectives on Social Psychology* (pp. 209–35). Boston: Allyn and Bacon.

Mortimer, Jeylan T., and Jon Lorence. 1979. "Work Experience and Occupational Value Socialization." *American Journal of Sociology* 84:1361–85.

Mottaz, Clifford J. 1981. "Some Determinants of Work Alienation." *Sociological Quarterly* 22:515–29.

Nietzsche, Friedrich. [1887] 1974. *The Gay Science*. (Walter Kaufmann, editor). New York: Random House.

Noble, David F. 1984. *Forces of Production*. New York: Knopf.

Ogburn, William F. 1957. "Cultural Lag as Theory." *Sociology and Social Research* 41(January–February):167–74.

Ogburn, William Fielding. 1964b. *On Culture and Social Change*. (Otis Dudley Duncan, editor). Chicago: University of Chicago Press.

Ong, Walter J. 1982. *Orality and Literacy*. London and New York: Methuen.

Orwell, George. 1949. *Nineteen Eighty-Four*. New York: Harcourt Brace.

Osgood, Charles, et al. 1967. *The Measurement of Meaning*. Urbana: University of Illinois Press.

Pappenheim, Fritz. 1959. *The Alienation of Modern Man*. New York: Monthly Review.

Pecotche, Carlos Bernardo Gonzalez (Raumsol). 1985. *Logosophical Exegesis*. Sao Paulo: Editora Logosophica.

Pecotche, Carlos Bernardo Gonzalez (Raumsol). 1986. *The Mechanism of Conscious Life*. Sao Paulo: Editora Logosophica.

Pecotche, Carlos Bernardo Gonzalez (Raumsol). 1991. *Deficiencies and Propensities of the Human Being*. Sao Paulo: Editora Logosophica.

Peirce, Charles S. 1955. *Philosophical Writings of Peirce*. New York: Dover.

Phillips, Bernard S. 1972. *Worlds of the Future: Exercises in the Sociological Imagination*. Columbus, OH: Charles E. Merrill.

Phillips, Bernard S. 1979. *Sociology: From Concepts to Practice*. New York: McGraw-Hill.

Phillips, Bernard S. 1980. "Paradigmatic Barriers to System Dynamics." *Proceedings of the International Conference on Cybernetics and Society, Cambridge, MA, October 8–10* (pp. 682–88). *Cambridge, MA: IEEE.*

Phillips, Bernard S. 1985. *Sociological Research Methods.* Homewood, IL: Dorsey.

Phillips, Bernard S. 1988. "Toward A Reflexive Sociology." *American Sociologist* 19(Summer):138–51.

Phillips, Bernard S. 1990. "Simmel, Individuality, and Fundamental Change." In Michael Kaern, Bernard S. Phillips and Robert S. Cohen (Eds.), *Georg Simmel and Contemporary Sociology.* Dordrecht/Boston/London: Kluwer.

Phillips, Bernard S., and Peter M. Senge. 1972. "Appendix: From Beaker Metaphor to a Dynamic Mathematical Model of Human Development." In *Worlds of the Future: Exercises in the Sociological Imagination.* Columbus, OH: Merrill.

Powell, Walter W., and Laurel Smith-Doerr. 1994. "Networks and Economic Life." In Neil J. Smelser and Richard Swedberg (Eds.), *Handbook of Economic Sociology* (pp. 368–402). Princeton, NJ: Princeton University Press.

Quine, W. V. O., and J. S. Ullian. 1970. *The Web of Belief.* New York: Random House.

Raushenbush, Hilmar S. 1969. *Man's Past: Man's Future.* New York: Delacorte.

Retzinger, Suzanne. 1991. *Violent Emotions: Shame and Rage in Marital Quarrels.* Newbury Park, CA: Sage.

Retzinger, Suzanne, and Thomas J. Scheff. 2000. "Emotion, Alienation, and Narratives: Resolving Intractable Conflicts." *Mediation Quarterly* (Fall):71–86.

Ricoeur, Paul. 1970. *Freud and Philosophy.* New Haven, CT: Yale University Press.

Ries, Lyn M. 1992. "Social Mobility." In Edgar F. Borgatta (Ed.), *Encyclopedia of Sociology* (pp. 1872–80). New York: Macmillan. 1992.

Roberts, Nancy, et al. 1983. *Introduction to Computer Simulation: The System Dynamics Approach.* Reading, MA: Addison-Wesley.

Rokeach, Milton. 1968. *Beliefs, Attitudes, and Values.* San Francisco: Jossey-Bass.

Rorty, R. 1987. "Science as Solidarity." In John Nelson (Ed.), *The Rhetoric of the Human Sciences.* Madison: University of Wisconsin Press.

Rosenthal, Robert. 1966. *Experimenter Effects in Behavioral Research.* New York: Appleton.

Rosenthal, Robert, and Lenore Jacobson. 1968. *Pygmalion in the Classroom.* New York: Holt.

Sapir, Edward. 1929. "The Status of Linguistics as a Science." *Language* 5:207–14.

Saposnek, Donald. 1983. *Mediating Child Custody Disputes.* San Francisco: Jossey-Bass.

Sayles, Marnie L. 1984. "Relative Deprivation and Collective Protest." *Sociological Inquiry* 449–65.

Scheff, Thomas J. 1966. *Being Mentally Ill: A Sociological Theory.* Chicago: Aldine.

Scheff, Thomas J. 1974. "The Labelling Theory of Mental Illness." *American Sociological Review* 39(June):444–52.

Scheff, Thomas J. 1990. *Microsociology: Discourse, Emotion, and Social Structure.* Chicago: University of Chicago Press.

Scheff, Thomas J. 1994. *Bloody Revenge: Emotions, Nationalism, and War.* Boulder, CO: Westview.

Scheff, Thomas J. 1997. *Emotions, the Social Bond, and Human Reality: Part/Whole Analysis.* Cambridge: Cambridge University Press.

Schneider, Louis. 1975. *The Sociological Way of Looking at the World*. New York: Mc-Graw-Hill.

Schutte, Jerald. 1977. *Everything You Wanted to Know About Elementary Statistics (but were afraid to ask)*. Englewood Cliffs, NJ: Prentice-Hall.

Scott, M. H., and S. M. Lyman. 1968. "Accounts." *American Sociological Review* 33:46–62.

Seeman, Melvin. 1959. "On the Meaning of Alienation." *American Sociological Review* 24:783–91.

Seeman, Melvin. 1972. "Alienation and Engagement." In Angus Campbell and Philip E. Converse (Eds.), *The Human Meaning of Social Change*. New York: Russell Sage.

Seeman, Melvin. 1975. "Alienation Studies." In Alex Inkeles, James Coleman, and Neil Smelser (Eds.), *Annual Review of Sociology* 1:91–123.

Seeman, Melvin. 1983. "Alienation Motifs in Contemporary Theorizing." *Social Psychology Quarterly* 46:171–84.

Shama, Simon. 1989. *Citizens: A Chronicle of the French Revolution*. New York: Knopf.

Shepard, Jon M. 1971. *Automation and Alienation*. Cambridge, MA: MIT Press.

Simmel, Georg. [1903] 1971. "Metropolis and Mental Life." In Donald N. Levine (Ed.), *On Individuality and Social Forms* (pp. 324–39). Chicago: University of Chicago Press.

Simmel, Georg. 1955. *The Web of Group Affiliations*. Glencoe, IL: Free Press.

Sjoberg, Gideon. 1965. *The Preindustrial City*. New York: Free Press.

Skocpol, Theda. 1979. *States and Social Revolutions*. Cambridge: Cambridge University Press.

Snow, David A. 1999. "1998 PSA Presidential Address: The Value of Sociology." *Sociological Perspectives* 42(1):1–22.

Snow, David A., and Pamela E. Oliver. 1995. "Social Movements and Collective Behavior." In Karen S. Cook, Gary Alan Fine, and James S. House (Eds.), *Sociological Perspectives on Social Psychology* (pp. 571–99). Boston: Allyn and Bacon.

Srole, Leo. 1956 "Social Integration and Certain Corollaries." *American Sociological Review* 21(December):709–16.

Stouffer, Samuel, et al. 1949. *The American Soldier*, Vol. 1. Princeton, NJ: Princeton University Press.

Sullivan, S. 1989. "Refining the Revolution." *Newsweek*, February 20.

Sztompka, Piotr. 1993. *The Sociology of Social Change*. Cambridge, MA: Blackwell.

Taylor, Stan. 1984. *Social Science and Revolutions*. London: Macmillan.

Thomas, W. I. 1923. *The Unadjusted Girl*. Boston: Little Brown.

Tilly, Charles. 1978. *From Mobilization to Revolution*. Reading, MA: Addison-Wesley.

Tocqueville, Alexis de. 1955. *The Old Regime and the French Revolution* (Stuart Gilbert, translator). New York: Doubleday.

Tumin, Melvin. 1953. "Some Principles of Stratification: A Critical Analysis." *American Sociological Review* 18(August):387–93.

Udy, Stanley H., Jr. 1959. "'Bureaucracy' and 'Rationality' in Weber's Organization Theory." *American Sociological Review* 24:591–95.

Vance, Jack. 1958. *The Languages of Pao*. New York: Daw.

Vidich, Arthur, and Joseph Bensman. 1960. *Small Town in Mass Society*. Garden City, NY: Doubleday.

Vygotsky, Lev S. [1934] 1965. *Thought and Language*. Cambridge, MA: MIT Press.

Wallace, Michael, and Arne L. Kalleberg. 1982. "Industrial Transformation and Decline of Craft." *American Sociological Review* 47:307–24.

Wallerstein, Immanuel. 1980. "The *Annales* School: The War on Two Fronts." *Annales of Scholarship* 3(Summer):85–91.

Wallerstein, Immanuel. 1991. "Beyond *Annales?*" *Radical History Review* 49(Winter):7–15.

Wallerstein, Immanuel. 1998. "Letters from the President, 1994–1998." International Sociological Association (unpublished).

Walsh, Edward J. 1982. "Prestige, Work Satisfaction, and Alienation." *Work and Occupations* 9:475–96.

Weber, Max. [1905] 1958. *The Protestant Ethic and the Spirit of Capitalism*. New York: Scribner's.

Weber, Max. 1958. "Class, Status, Party." In *From Max Weber* (pp. 180–95, H. H. Gerth and C. Wright Mills, translators). London: Oxford University Press.

Weber, Max. 1964. *The Theory of Social and Economic Organization*. New York: Free Press.

Whorf, Benjamin. 1963. *Language, Thought, and Reality*. Cambridge, MA: MIT Press.

Willer, David. 1967. *Scientific Sociology*. Englewood Cliffs, NJ: Prentice-Hall.

Willer, David, and Murray Webster, Jr. 1970. "Theoretical Concepts and Observables." *American Sociological Review* 35(August):748–57.

Williams, Robin M., Jr. 1970. "Major Value Orientations in America." In *American Society*, 3rd ed. (pp. 452–500). New York: Knopf.

Williams, Robin M., Jr. 1992. "American Society." In Edgar F. Borgatta (Ed.), *Encyclopedia of Sociology*, Vol. 1 (pp. 81–87). New York: Macmillan.

Wilson, Edward O. 1998. "Back from Chaos." *Atlantic Monthly* (March):41–62.

Wilson, William Julius. 1987. *The Truly Disadvantaged: The Inner City, the Underclass, and Public Policy*. Chicago: University of Chicago Press.

Wrong, Dennis H. 1961. "The Oversocialized Conception of Man in Modern Sociology." *American Sociological Review* 26(April):183–93.

Index

Action, 197, 201
Addiction, 24, 26, 29, 32–38, 72, 108, 127–128, 136, 154–155, 163, 170, 172, 176, 179, 192–193, 197–200, 217
Alienation, xii, 4, 7, 13, 17, 24, 28–29, 32–38, 48, 64–65, 72–74, 79–80, 88–91, 94–97, 100–101, 105–106, 108, 112–113, 118, 123–124, 127–128, 136, 147–149, 154–155, 163, 172, 176, 179, 191–193, 197–200, 204–205, 217–218, 220, 225, 227
Anomie, xii, 6–8, 13, 17, 24, 28, 32, 34, 36–37, 48, 65, 72–74, 79–82, 84–88, 91, 97, 100–101, 105–106, 112–113, 118, 123–124, 127–128, 136, 148, 154, 172, 176, 193, 197–198, 217, 219
Apel, Karl, 29

Bacon, Francis, 45–46, 200
Bailey, F. G., 70
Ball-Rokeach, Sandra J., 183
Banks, Jane, 104
Bell, Wendell, 88, 95
Bensman, Joseph, viii, 115–116
Bergesen, Albert, 98
Biological structure, 24–25, 42–43, 65, 87–89, 118, 197, 201, 220
Blalock, Hubert M., Jr., 175
Blau, Peter M., 62
Blauner, Robert, 95
Blumer, Herbert, 10–11, 13, 63, 70
Bondurant, Joan V., 139–143, 145, 204
Bowles, Samuel, 91
Braverman, Harry, 95
Brinton, Clarence Crane, 132–133
Brown, Donald, 137
Bunuel, Luis, 168–169
Bureaucracy, 12–14, 16, 24–26, 28, 32–33, 35–36, 44, 65, 74, 80, 93, 97, 119, 123–124, 127–128, 135–138, 154, 163, 166, 177, 184, 192, 197–199, 204, 206, 209
  Bureaucratic scientific method, viii–ix,
xi–xiii, 1–3, 5, 10–39, 41–42, 50–72, 176–182, 188–190
  Bureaucratic worldview, vii–x, xii–xiii, 1, 6–7, 12–39, 41–59, 72, 79–159, 161–194
Buroway, Michael, 95

Causal-loop analysis, xii, 32–39, 125–129, 188–189, 202–206
Chapin, F. Stuart, 88
Chasin, Barbara H., 4
Clinton, William Jefferson, 195–196, 226
Collins, Randall, 65, 97
Comte, Auguste, vii–vii, xiv, 64, 74, 157, 216
Conformity, 24–25, 32, 34, 46, 83, 87, 109, 119–120, 122–124, 137–138, 145, 147, 153–154, 158, 163, 166, 169, 172, 174, 176–177, 185, 192–193, 197–198, 204–205, 207, 209, 217
Constas, Helen, 199
Cook, Carol S., 62
Cooley, Charles H., 151
Cronbach, Lee J., 29
Crutchfield, Robert D., 85
Culture, x–xi, xiii, 1, 2, 5, 8, 11, 13–18, 20, 24–29, 32–33, 35–38, 41–42, 44–45, 48–51, 54–55, 59–67, 69–71, 73–74, 76, 79–88, 91–93, 96–101, 103–107, 109, 111–113, 119–124, 127–129, 131, 136–138, 144–145, 159, 163, 165–167, 169–179, 181–184, 186–188, 190–193, 195–197, 199–200, 202, 204–206, 209, 214–219, 220–221, 223–224, 226–227

Dahrendorf, Ralf, 97
Darwin, Charles, 151
Davies, Kingsley, 132–134
Davis, Kingsley, 97
Definition of a problem, vii–viii, 4–7, 18–21, 30–35, 52–54, 64, 66–72, 79–113, 129–159, 167–168, 175–182